SNORKEL HAWAI'I

The Big Island

W9-AAI-471

Guide to the Underwater World • Judy and Mel Malinowski

Snorkel Hawai'i: The Big Island
Guide to the Underwater World
1st Edition

Published by: Indigo Publications
                 920 Los Robles Avenue
                 Palo Alto, CA 94306 USA

                 SAN 298-9921
                 Publisher's symbol: Indigo CA

First Published: 1996
Printed by Publishers ExpressPress, Ladysmith, WI 54848 USA

About the cover:

Camille Young painted this lovely Picasso triggerfish in watercolor especially for our cover. A graduate of the University of Hawai'i at O'ahu, she now lives in Moraga, California.

Dave Barry is renowned for his humorous essays and books. His love of the underwater world brings a special eloquence to these passages.

Mahalo to the kind residents of The Big Island for keeping the Aloha tradition alive, The Hawai'ian Visitors Bureau, Marta Jorasch, Erland Patterson, Dave Barry, and many others.

Every effort has been made to provide accurate and reliable information and advice in this book. Venturing out into the ocean has inherent, ever changing risks which must be evaluated in each situation by the persons involved. The authors and publishers are not responsible for any inconvenience, loss, or injury sustained by our readers during their travels. Travel, swim and snorkel safely – when in doubt, err on the side of caution.

Quotes from "Blub Story", Tropic Magazine, © Dave Barry 1989.

ISBN 0-9646680-0-9

Library of Congress Catalog Card Number: 96-94357

CONTENTS

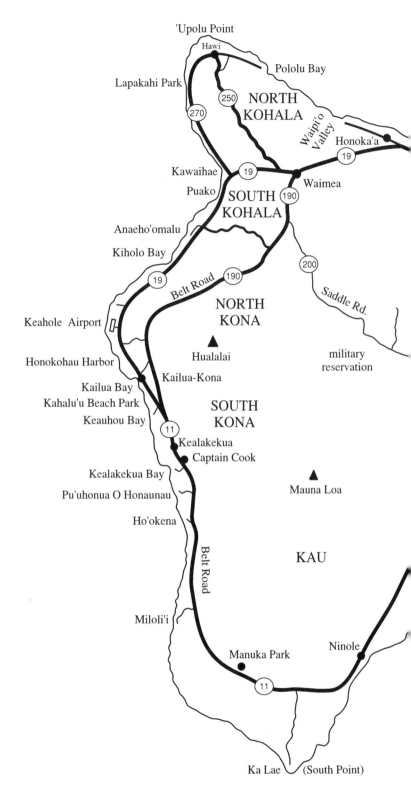

THE BIG ISLAND

(Also see Snorkeling Site Index Map on page 36)

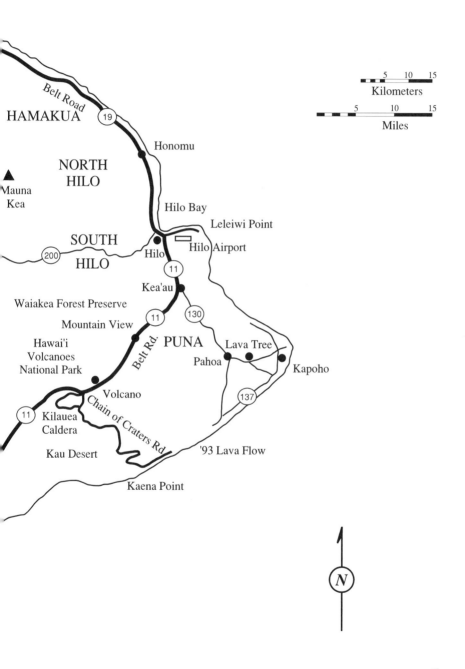

Belt Road
HAMAKUA 19

Honomu

NORTH HILO

▲ Mauna Kea

Hilo Bay

Leleiwi Point

SOUTH 200
HILO

Hilo
Hilo Airport

11

Kea'au

Waiakea Forest Preserve

Mountain View 11 130

Hawai'i Volcanoes National Park

PUNA

Lava Tree

Pahoa

Kapoho

Volcano 137

11 Kilauea Caldera

Chain of Craters Rd

Kau Desert

'93 Lava Flow

Kaena Point

Belt Rd.

5   10   15
Kilometers
5   10   15
Miles

N

5

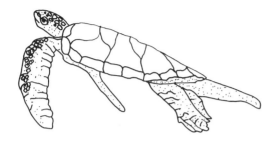

to Marta Jorasch,
our island girl

WHY SNORKEL THE BIG ISLAND?

We came from the sea; our blood is salty, and our tears. Many years have passed since our primitive ancestors left the sea, yet the water still beckons us in our dreams. Snorkeling allows us to follow those dreams to their source, and enjoy the most colorful show on earth.

When you don that mask and snorkel, and gently place your face toward the sea, you've entered another planet – one completely unlike home. Everything works differently here under the sea. It's graceful, soft and inviting, with a dazzling array of color and whimsical life forms.

Scuba, with its elaborate training and heavy, expensive equipment, is one way to enter this world. Those of us who love what we see using scuba still don't love the gear. Snorkeling is a lighter and easier way, a family sport accessible to nearly all ages and abilities. Once you've arrived in Hawai'i, it's a remarkable bargain – the best in the islands.

An active vacation is memorable for adventure as well as relaxation. Hassles and missteps finding out where to go can raise your blood pressure, and waste your time. Our research will help you quickly locate appropriate sites that fit your interests and abilities, saving your valuable vacation hours.

Snorkeling sites in Hawai'i are trickier than some calmer parts of the world, so it's best to get good advice before heading out. Everyone's had their share of unpleasant experiences due to vague directions, as well as outdated or inaccurate information. We have created the *Snorkel Hawai'i* series as that savvy snorkeling buddy everyone needs. We hope you'll enjoy the personal stories we've included; see *About the Authors* on page 126 if you want to know a little more about us.

We have personally snorkeled all the major sites listed. Few of these sites are well-marked. The challenge lies in knowing how to find them quickly, as well as how to enter and exit, and where to snorkel, so you'll have a safe and rewarding experience. Our detailed maps and instructions will ease the uncertainty, saving you time and effort.

The Big Island offers some of the clearest water and finest snorkeling sites in all of the Hawai'ian Islands. Less developed than the more crowded islands of Maui and O'ahu, it retains much of the Aloha spirit even today. Most residents of the Big Island are friendly, relaxed and willing to talk and share their love of this remarkable place. Try to visit it at least once in your life, and by all means don't miss the underwater world. Aloha!

–Judy and Mel Malinowski

- easy
- relaxing
- fun
- floating on the surface of the sea
- breathing without effort through a tube
- peering into the water world through a mask
- open to anyone of any age, size, shape or color

SPOTTED BOXFISH

Who was the first snorkeler? As the fossil records include few petrified snorkels, we are free to speculate.

Among larger creatures, perhaps elephants are the pioneers and current champions, as they have known how to "snorkel" for countless generations. Once in a blue moon, you may see an elephant herd heading out to do lunch on an island off the coast of Tanzania, paddling along with their trunks held high. No one knows whether the hefty pachyderms enjoy the fish-watching, but you can bet a big liquid chuckle reverberates through the ranks of reef fish in the vicinity as the parade goes by.

As evolution continued, perhaps a clever homo sapiens saved his furry brow by hiding underwater from pursuers, breathing through a hollow reed. Surviving to propagate his brainy kind, he founded a dynasty of snorkelers. Perhaps he actually liked the peaceful atmosphere down there, and a new sport was born. Masks, however, came much later, so the fish looked a little fuzzy.

Some of our readers may grumble that snorkeling is not a real sport: no rules, no score, no competition, scarcely aerobic, with hardly any equipment or clothing. We say to them: lighten up, you're on vacation!! Go for a long run later. Diehard competitors can create their own competition by counting the species they've seen or trying to spot the biggest or the most in one day.

BASICS

To snorkel, you need only two things:

Snorkel
: Saves lifting your head once a minute, wasting energy and disturbing the fish.

Mask
: While you can see (poorly) without one, it keeps the water out of your eyes and lets you see clearly.

Rent them inexpensively at many local shops, or buy them if you prefer. It's all the back-to-basics folks need to snorkel in calm warm water, where there aren't any currents or hazards, such as your hotel pool or your hot tub.

Savvy snorkelers often add a few things to the list, based on years of experience, such as:

Swimsuit
: Required by law in many localities, possibly excluding Brazil. Added benefit: saves you from an occasional all-body sunburn.

Fins
: Good if you want to swim with ease and speed, like a fish. Saves energy. A must in Hawai'i, due to occasional strong currents. They protect your tender feet, too.

T-shirt
: Simple way to avoid or minimize sunburn on your back. Available everywhere.

Sunscreen
: To slather on your tender exposed backside skin – legs, neck, backs of your arms. Not optional in the tropics for light-skinned snorkelers.

Wetsuit
: For some, the Hawai'ian waters seem a bit chilly – not exactly pool-warm. Wetsuits range from simple T-shirt-like tops to full suits. Worth considering.

You're almost ready to get wet. But wait!

You want to know even more technical detail? Just ahead, we'll go into enough detail to satisfy your deepest inner technical cravings. A major equipment shopping list if that's what you want. Every sport has an equipment list – it's what keeps sporting goods stores in business, and your garage shelves full.

GEAR SELECTION

Good snorkeling gear enables you to pay attention to the fish, instead of uncomfortable distractions. Poor equipment will make you suffer in little ways, from pressure headaches from a too-tight mask, to blisters on your feet from ill-fitting fins. Consider well before buying.

SNORKEL

Snorkels can be quite cheap. Be prepared to pony up $15 or more if you want them to last awhile.

You'll appreciate a comfortable mouthpiece if you plan to snorkel for long. Watch out for sharp edges – a good mouthpiece is smooth and chewy-soft.

Several new high tech models have been designed to minimize water coming down the tube from chop or an occasional swell overtopping you. We looked at these with mild skepticism until a choppy trip had us coughing and clearing our snorkels every third breath.

We took our new snorkel out to the hot tub, and tried pouring buckets of water down the tube. The snorkeler didn't even notice! Our verdict is: the new technology ones really work as advertised. (Avoid the old float ball at the top versions, though).

We use the US Divers Impulse snorkel: ($35+), but others may be equally effective. An acquaintance uses and recommends a Dacor model with corrugated, flexible neck.

A bottom purge valve makes blowing out water smooth and easy, on those occasional cases when it is required. It's well worth the higher price if you snorkel in choppy water or like to surface dive.

We found that turning the "orthodontic mouthpiece" upside down on our Impulse snorkel made it more comfortable. The mouthpiece is available in two sizes, so be sure to get the one that fits your mouth. There is a company that will custom-make a mouthpiece just for you if you have a hard-to-fit mouth. Ask at your local dive shop.

You may have several colors to choose among. Although we're not especially fond of neon, this helps to be spotted more easily in the water, since only the top of your snorkel is very visible. Bright colors could help you find dropped pieces underwater, too.

SNORKEL HOLDER

This little guy holds your snorkel to your mask strap, so you don't keep dipping it in the sea. The standard is a simple figure 8 double loop that pulls over the snorkel tube, wraps around your mask strap, and then back over the tube. If you are replacing one, be sure to get the right size for your tube diameter, or one that converts. Prices range from $.50 to $3. A hefty rubber band will work passably in a pinch.

An alternative high tech holder has a slot that allows the snorkel to be adjusted easily. It slides rather than having to be tugged, so is a bit easier on long hair. These come with some higher-end snorkels.

The standard scuba snorkel positioning is to your left side – might as well get used to it, as you may dive eventually.

MASK

Nothing can color your snorkeling experience more than an ill-fitting mask. (Unless, of course, you get that all-body sunburn mentioned earlier). Don't settle for painful or leaky masks! If it hurts, it's wrong.

In this case, 'pain, no gain' applies.

Simple ones can cost as little as $10. Top quality ones run upwards of $60, into the low three figures in some cases, unless on sale. It's probably better to start out with a rental mask, paying a bit extra for the better quality models. As you build more experience, you'll be in a better position to evaluate a mask before you shell out a lot of money.

You need a good fit to your particular face geometry. Shops often tell you to place the mask on your face (without the strap) and breathe in. If the mask will stay in place, then they say you have found a good fit. However, nearly all masks will stay on my face under this test, and yet some leak later!

Look for soft edges and a mask that conforms to your face even before drawing in your breath. There's a great deal of variance in where a mask rests on your face and how soft it feels, so compare very carefully. Look for soft and comfortable, unless you especially like having pressure headaches, or looking like a very large octopus glommed on to your face.

Good visibility is certainly important. We happen to prefer masks with glass side panels. You can't really see fish well through the side, but your peripheral vision is improved, so you can keep track of a companion without lifting your head out of the water so often.

Lack of 20-20 vision needn't cut into your viewing pleasure, but it does require a little more effort during equipment selection. Those who wear contact lenses can use them within their masks, taking on the risk that they'll swish out and float softly and invisibly down to the sea bed, perhaps to be found by a fossil hunter in the distant future, but certainly not by you. Use the disposable kind.

Vision-correcting lens are available for many masks, in 1/2 diopter increments. Unless you use contacts, search for a correctable mask. It's a real shame to finally get to those great fish, and then find them all blurry, like a TV with no antenna.

Low volume masks

There are a variety of shapes and forms of mask lens designs. The inexpensive masks tend to have one large flat front glass. They're OK if the skirt of the mask fits you – although they often are a bit stiff and uncomfortable. They also tend to be far out from your face, with a big air space. As you go up in price, the lenses tend to get smaller and closer to your eyes, as preferred by divers. Why is this so?

There is a good (scuba) reason for this. These are called 'low volume' masks. They contain less airspace, and so require less effort to clear when water gets in. They also press less against your face as you go deeper, (unless you blow higher pressure air in through your nose) and hence are more comfortable when diving.

For a snorkeler, this seems of little importance at first glance. It still should be considered as you select your mask, however, for possible future benefits.

Many snorkelers go on to do some surface diving, as well as snuba or scuba diving, and low volume is an advantage for the reasons mentioned above. When you dive down even 10', the water pressure is considerable. At 32', the air in your lungs and mask is compressed to half its volume, and unless you remember to blow some air into your mask through your nose, the pressure on your face can be most uncomfortable!

Likewise, if your mask is flooded, which does happen occasion-ally, it is easier to clear out the water from a low volume mask.

If the mask you prefer doesn't offer standard correcting lenses, custom prescription lenses can be fitted to almost any mask. This costs more and takes longer, but does produce the most accurate result. Bifocals are available. If you do opt for these expensive add-ons, we suggest that you water test the mask for comfort and leakage before making this extra investment.

Mustaches create a mask leakage problem. As I choose to have a mustache, I have coped with this my entire adult life, excluding one amusing trip.

On a plane to Anguilla, I thought I would neatly solve the leak problem and give Judy a real surprise. I made a routine trip to the bathroom, while hiding a razor in my pocket. Off came the mustache, and then I returned, and quietly sat back down. I turned to my traveling companion, and she just about fell out of her seat! This was *not* the man she married, and she had the odd experience of feeling like she was vacationing with a stranger. The verdict: she preferred her familiar old companion. Mustache leaks are better than marital discord.

There are those who use a petroleum jelly, such as Vaseline, to make the seal. That doesn't appeal to me, going in and out of the water twice or more a day. It does help to choose a mask that rests high over the mouth and perhaps trim the top 1/8" or so of the center mustache, if it sticks up. Hair breaks the seal and allows water to seep into the mask slowly, so you'll still have to clear the mask occasionally.

Anyone with a leaky mask may prefer having a purge valve. There are some clever higher-end purge valve masks. The conventional wisdom in scuba is that purge valves are an unnecessary weak point. Nevertheless, there are experienced divers who use them regularly without having any problems.

A purge valve can quickly and cheaply be added to some masks, by some dive shops (if you have a bit of space just under the nose), but it is a bit less fail safe for diving. You can also compensate by tightening the mask, but that's not particularly comfortable and leaves a distinct mask mark on your face at the end of the snorkel trip.

ELEGANT CORIS

13

MASK STRAP

The strap that comes with the mask is generally fine, but if you have your own mask and want it to slide on more easily, there's a comfortable strap available with adjustment by velcro. The back is made of wetsuit material, stretchy and soft. They supposedly will float if your mask should come off, but ours don't. A big benefit: no more hair tangles in the strap! Cost: about $12 in dive shops, but they're not always easy to find. Since we get in and out so often, we happen to prefer this one to a rubber strap, but it's a convenience for the frequent snorkeler, rather than a necessity.

FINS

The simplest fins are basic (usually black) enclosed foot fins. These are one-piece molded rubber, and slip right on to your bare feet. For warm water, basic snorkeling, these inexpensive fins are fine. We own several kinds of fins, and still often choose the one-piece rubber foot fins for lightness and easy packing. They seem to last forever. ($15–$25)

Why look further? For special uses, and higher performance. We recently performance tested three sets of fins, doing timed swims over a measured course. The basic fins discussed above went first. A set of fairly expensive, but rather soft, flexible strap-on fins cut the swim time by 20%, while ultra long, stiff-bladed foot-mount Cressi fins cut it by 40%! These long surface diving fins are, however, a little long and awkward to use for most surface snorkeling or even scuba.

There is a vigorous debate going on among divers about the merits of flexible fin blades versus stiff blades. We've tested both, and our opinion is that the most efficient are light, thin, stiff blades, hands down. We have a pair of top quality, but soft blade fins we'd just love to sell to you, cheap, if you are of a different opinion.

You're better off with a medium blade size foot fin for most snorkeling. Large diving fins are awkward for snorkeling, and require more leg strength than most non-athletes possess. The big diving fins do come in numerous shapes and colors, which some people are convinced will make them faster or perhaps more attractive. Speed is not the main aim of snorkeling, but has its uses.

Faster fins do enable you to cover more territory, and they also serve as excellent insurance in case you wander into a strong current. Unless it's absolutely certain that no current can carry you away, ALWAYS WEAR FINS!

As you look at more advanced fins, they split into two attachment methods, with pros and cons to each type. We own both, and pick the best for a particular situation. Some models are available in both attachment styles.

ENCLOSED FOOT   Your bare foot slides into a stretchy, integral molded rubber 'shoe'.

Advantages   This gives the lightest, most streamlined and fish-like fit. It probably is the most efficient at transmitting your muscle power to the blade. We like these best when booties are not required for warmth or safety.

Disadvantages   The fins must be closely fitted to your particular foot size. Some models may rub your skin wrong, creating blisters. If you have to hike in to the entry site, you need separate shoes. This may preclude entering at one spot, and exiting elsewhere. If you hike over rough ground (lava, for example) to get to your entry point, or the entry is over sharp coral or other hazards, these may not be the best choice.

STRAP-ON   Made for use with booties.

Advantages   Makes rough surface entry easy – just hike to the entry point, head on into the water holding your fins in hand, lay back and pull on your fins. Exiting is just as easy. The bootie cushions your foot, making blisters unlikely. Widely used for scuba.

Disadvantages   Less streamlined. The bootie makes your foot float up, so you may have trouble keeping your fins from breaking the surface at times. Some divers use ankle weights to counter this, but they can tire you, and slow you down.

REEF SHOES OR BOOTIES

Walking with bare feet on lava or coral can shred your feet in a quick minute. There are fine reef shoes available that are happy in or out of the water. These are primarily for getting there, or wading around, as they don't really work that well with strap-on fins. For the sake of the reef, don't actually walk on a reef with them, as each step kills hundreds of the little animals that make up the living reef.

Zip-on booties are widely used by divers, and allow use of strap-on fins. They do float your feet up some, a minor annoyance for snorkelers. Some divers use ankle weights to counter this.

Judy prefers her small enclosed foot fins, so she carries along cheap old shoes or reef shoes to cross a rough area, and then just leaves them at the entry point. Use old, grungy ones no one would want to steal. In Hawai'i lots of shops and markets sell reasonably priced versions of reef shoes, which are very handy if you like to explore.

KEEPING TIME

One easy-to-forget item: a water-resistant watch. This needn't be expensive, and is very useful. It's essential for pacing yourself, and keeping track of your sun exposure time. We prefer a slim, analog watch with a nice clean uncluttered face and easy to read numbers. Inexpensive simple Timex water-resistant watches have given us good performance, up until it's time to change the battery!

"Water resistant" alone usually means that a little rain won't wreck the watch, but immersion in water may. When "to 50 meters" is added, it denotes added water-resistance; but the dynamic pressures from swimming increase the pressure, so choose 50 meters or greater rating to be safe even when snorkeling. It does help to not press any of the buttons underwater.

Don't take a 50 meter watch scuba diving, though – that requires the 100-200 meter models. Hawai'ian time is two hours earlier than Pacific Standard Time, or three hours earlier than Pacific Daylight Time (as Hawai'i doesn't observe Daylight Savings Time).

ORNATE BUTTERFLYFISH

BODY SUITS

There are a variety of all-body suits that protect you from sun exposure and light abrasion, but provide little warmth. They are made from various synthetic fabrics, lycra and nylon being common. They cost less than wetsuits, and are light and easy to pack. These are more effective at warding off sunburn than T-shirts, and are also good for midday windsurfing or sailing. We often carry ours in case the water and weather are especially warm, but we'd prefer not getting nipped by bothersome jellyfish or little floating bits of hydroids that occasionally show up in numbers.

WETSUIT

Water temperature on the surface varies from a low of about 72° F in March to a high of about 80° in September. If you happen to be slender, no longer young and from a moderate climate, this can seem cold. Sheltered bays and tidepools can be a bit warmer, while deeper water can be surprisingly cold. We've snorkeled in March when we swore it was not above 65° off Kaua'i.

Regardless of the exact temperature, the water is cooler than your body. With normal exertion, your body still cools, bit by bit. After awhile, perhaps 30-45 minutes, you start feeling a little chilly. Later, you begin shivering, and then eventually, hypothermia begins.

A wetsuit isn't necessary in Hawai'i, but sure makes being in the water a lot more fun for many who are slim or who have lower metabolism. We like to snorkel for two hours or more, and a thin wetsuit protects us from the sun, and keeps us warm and comfortable as well.

Off the rack suits are a bargain, and fit many folks. We have seen Costco carrying perfectly adequate shortie (short sleeve arms and above-the-knee legs) suits for under $60. Watch out for a snug fit at neck, arms and legs – if your suit is loose there, water will flow in and out, making you cold. If you have big feet and small ankles, get zippers on the legs if possible if buying a full length suit.

We are both tall and skinny, so off-the-shelf suits don't fit well. Our standby suit is a light, full body custom-made suit, with 3mm body, and 2mm arms and legs. Judy had zippers installed on both arms and legs. After we got these suits, our pleasure level while snorkeling in Hawai'i went way up. They also suffice for warm-water diving.

Wetsuit wearers also get added *range*. You can stay in the water without hypothermia for many hours. This could be comforting in the unlikely event that some strong current sweeps you off towards Fiji.

SWIM CAP

If you have trouble with long hair tangling in your gear while snorkeling, a Speedo swim cap can help this as well. A cap is particularly useful if you're entering the water multiple times per day, since wet, salty hair has a perverse way of weaving itself into the buckles and straps of your mask.

SNORKELING VEST

It is possible to buy inflatable vests made for snorkeling. Some guidebooks and stores promote them as virtually essential. We've even been on excursion boats that require all snorkelers to wear one.

Use your own judgement, with prudence erring on the side of caution, of course. Vests are hardly necessary in salt water for most people, but useful if you can't swim a lick or won't be willing to try this sport without it. There is a possible safety edge for kids or older folks. If you do get a vest, you can give it to another beginner after you get used to snorkeling. You'll have discovered that it takes almost no effort to float flat in the water while breathing through a snorkel!

If you want extra flotation, consider using a light wetsuit instead – it simultaneously gives you a little more buoyancy, sun and critter protection, and warmth, too!

SURFACE DIVING GEAR

For surface diving, bigger fins help your range. Those surreal-looking Cressi fins that seem about three feet long will take you down so fast you'll be amazed.

Be careful down there, though. Periodically, you read about yet another expert surface-diver who passed the oxygen deprivation edge, blacked out and drowned. If you like living more than danger, don't push your limits too far. With a little good judgement, though, surface diving can be both fun and safe.

A long-fin alternative is to use a weight belt with from 2-4 pounds – just enough to help you get under the surface without using up all your energy. As you descend, you become neutrally buoyant at about 15-20' so you don't have to fight popping up. Of course, the sword cuts two ways, as you must swim up under your own power in time to breathe!

MOORISH IDOL

Motion sickness

Motion sickness ("seasickness or carsickness") is a minor inner ear disorder which can really cut into your pleasure on the water, on long, curvy road trips, or in choppy air. Fortunately, motion sickness is quite controllable these days. All it takes is a little advance planning to turn a potentially miserable experience into a normal, fun one.

Mel can get seasick just by imagining a rocking boat, so he's tried just about every remedy personally. What really works?

Forget the wrist pressure-point bands – they don't do the job for anyone we've met. Put them in the closet along with your ultrasonic pest repeller.

A time-release ear patch has been on the market, called Scopolamine, that is effective, but with possible side effects we'd rather not experience. For the time being, we avoid it.

The most effective remedy we've found by far is Meclizine, available by prescription only. It works perfectly for Mel, with no noticeable side effects. We learned about it when Jon Carroll, a favorite local columnist, reported that it had sufficed for him in 15-25' swells on the way to Antarctica. If it does the job there, it should handle most snorkeling excursions.

An alternative that works pretty well is Benadryl, usually used as a decongestant, available over the counter. It is also effective against motion sickness. Mel discovered this accidentally when he used Benadryl before a rough night dive, and felt great.

Use these medicines carefully, and only after consulting your doctor. In some cases, you must avoid alcohol or other drugs due to drug interaction problems. Take care driving or scuba diving, too, as some medicines can produce drowsiness.

19

INTO THE WATER

GETTING STARTED

Now you've assembled a nice collection of snorkel gear. You're ready to go! On a sunny tropical morning you have walked down to the water's edge. Little one-foot waves slap the sand lightly, while a soft warm breeze takes the edge off the intensity of the climbing sun. It's a great day to be alive, and out in the water.

Since you're going snorkeling, you've refrained from applying suntan lotion on your face. You sure don't want it washing into your eyes, to make them burn and water. You've worn a nice big hat instead. You applied lotion to your back before you left, so it has time to become effective, and you washed off your hands and rinsed them well.

CHECKING CONDITIONS

Take it nice and slow, one thing at a time. Sit down and watch the waves for a few minutes. Look for their patterns, how big the biggest waves are, and how far they wash up on the beach. When you see the pattern, you're ready to go. Set your gear down back beyond where the furthest watermarks are on the sand. You don't want that seventh wave to sweep your gear away!

GEARING UP

THREADFIN BUTTERFLYFISH

Take your mask in hand, and defog it. You treat the lens so that water vapor from your nose, or water leakage, won't bead up on your mask lens and spoil your view. There are two main chemical compounds for this, one old as us, and the other a product of modern chemistry.

The classic solution is: SPIT. Spit on the inside of your dry mask lens, and rub it all around with your sunscreen-free finger. Step into the water, just out beyond the stirred up sand, and dip up a mask full of clear saltwater. Thoroughly rub and rinse off that spit, and dump the mask. Voila, a mask that is fog-resistant for an average snorkel!

If you spit and polish, and still have fogging problems, there are several possible causes. Your mask may be gooped up with cosmetics, dried on saltwater residue, or whatever other goos may be out there. A good cleaning with toothpaste is in order (*see Caring for Your Gear, page 24*). It's possible that you didn't actually wet all the surface with spit, perhaps because there were drops of water left on the lens.

Of course, you may be spit-deficient, which government studies have determined occurs in 14% of the snorkeling population. Tough break. In that case, or if you just feel funny about spitting in your mask, you can use no-fog solution.

No-fog solution for masks actually *does* work even better than spit. It comes in small, handy, inexpensive bottles that seem to last forever, because you use only a few drops at a time.

Our favorite trick is to pre-apply no-fog to the dry masks an hour or more ahead, and let it dry on. When you get to the water, just rinse out thoroughly, and you get even better results!

GETTING COMFORTABLE

After you rinse your mask, try its fit. Adjust the mask strap and snorkel until they're quite comfortable. Hold the snorkel in your mouth without tightening your jaws. It can be quite loose and will not fall out.

When you like the fit, pull it down over your head so it rests on your forehead or chest. In some scuba circles, putting your mask up on your forehead is a signal of distress, so it will make some divers nervous if you do it. We don't find this rule convincing, at least for snorkelers. Your forehead is about the most practical mask storage place. Putting your mask on too early can cause it to fog from your exertions.

GETTING WET

Retrieve your fins and walk back in the water, watching the waves carefully. NEVER turn your back on the ocean, lest a rogue wave sneak up on you and whack you good.

If the bottom is sandy smooth, wade on out until you're about waist deep. Pull your mask on, making sure you remove any stray hair from under the skirt. Position the snorkel in your mouth, and start breathing. You can practice this back in your room, or even at a pool or in the hot tub if you like acting wild and crazy. Don't snorkel under the influence, though!

Duck down in the water so you're floating, and pull on your flippers just like sneakers. Make a smooth turn to your stomach, pause to float and relax until you're comfortable, and you're off! Flip those fins, and you have begun your re-entry into the sea.

As you float, practice steady, even breathing through the snorkel. Breathe slowly and deeply. People sometimes tense up at first and take short breaths. When this happens, you're only getting stale air from the snorkel rather than lots of fresh air from outside. If you ever feel tired or out of breath, don't take off your mask and snorkel – just stop for several minutes, float, breathe easy, and relax.

After you're quite comfortable breathing this way, check how your mask is doing. Make sure it isn't leaking. Have your buddy check it out. Adjust the strap if needed. And keep adjusting until it's just right. Slide your snorkel strap to a comfortable position, with the tube pointing about straight up as you float looking down at about a 30° angle. This is a good time to remove any sand that got in your flippers.

Swimming while snorkeling is extremely easy once you've relaxed (*see Snorkeling is easier than swimming, page 24*). No arms are required. What works best is to hold your arms straight and smooth back along your sides, keep your legs fairly straight and kick those fins slowly without bending your knees. Any swimming technique will work, of course, but some are more tiring. Practice using the least amount of energy just to make sure you can do it. Once you learn how to snorkel the easy way, you can use all the power you like touring large areas as if you were a migrating whale.

CLEARING YOUR MASK

Eventually you will need to practice clearing your mask of water that leaks into it. The scuba method: Take a deep breath, then tip your head up, but with the mask still just under the surface. Press your palm to the top of the mask against your forehead, or hold your fingers on the top of the mask while lightly easing pressure on the bottom of the mask, and exhale through your nose. This forces the water out the bottom of the mask. You may need to do this more than once to get all the water out.

Lifting your head out of the water and releasing the seal under your nose will work too, but is frowned on by dive instructors, because it won't work under water. They would hate to see you develop bad habits if there's any chance you'll take up diving later. It also uses up more energy, and is harder in choppy conditions.

TAKING IT EASY

Relax and try not to push yourself too hard. Experienced snorkelers may urge you on faster than you're comfortable because they've forgotten how it feels to get started. As your experience builds, you'll find it easy, too.

It's a bit like learning to drive a car. Remember how even a parking lot seemed like a challenge? It helps to practice your beginning snorkeling in a calm easy place – with a patient teacher. With a little persistence, you'll soon overcome your fears and be ready. If you're having a slow start, try snorkeling in the shallow end of your hotel pool.

KNOWING YOUR LIMITS

Have you heard the old saloon saying: "Don't let your mouth write checks that your body can't cover"?

Let's paraphrase this as "Don't let your ego take you places your body can't get you back from." Consider carefully how well-conditioned your legs are, so you'll have enough reserve to be able to make it back home, and then some, in case of emergency.

Pacing

When you're having a good time, it's easy to forget and overextend yourself. That next rocky point beckons, and then a pretty spot beyond that. Pretty soon, you're miles from home, and getting cold and tired. Getting overly tired can contribute to poor judgement in critical situations, making you more vulnerable to injury. Why risk turning your great snorkeling experience into a pain? Learn your limits, and how to pace yourself.

Our favorite: If we plan on a 1 hour snorkel, we watch the time, and start heading back when we've swum 30 minutes. If the currents could run against us on the way back, we allow extra time/energy.

Snorkeling is easier than swimming

Some folks never learn to snorkel because they're not confident as swimmers. This is an unnecessary loss, because snorkeling is actually easier than swimming. We have maintained this to friends for years, and noted their doubtful looks. Recently, we came across a program in California that actually uses snorkeling as a tool to help teach swimming!

The Transpersonal Swimming Institute in Berkeley, CA specializes in the teaching of adults who are afraid of the water. Local heated pools are used all year. But the warm, salty and buoyant ocean is the best pool of all.

Melon Dash, Director of TSI, takes groups of her students to Kona on the Big Island 5-7 times a year. They begin by floating comfortably in the warm, salty Hawai'ian water. At their own pace, they gradually learn to snorkel, and then to swim at ease.

"We have found that people cannot learn what to do with their arms and legs while they are afraid that they might not live."

With a steady air supply, not having to worry about breathing in water accidentally, they can relax and learn the arm and leg movements at ease. Happily, they soon discover there's nothing complicated about it!

In calm conditions, with warm water, there need be no age limits and few physical limits for snorkeling. Some lap swimmers prefer to swim with mask and snorkel, as they get a full, easy air supply that way, and no chlorine in their eyes.

Melon Dash is a swimmer and teacher who guarantees results. Her 10-year program has been so successful that she plans to add classes on other islands, including Maui and O'ahu.

Contact:

Transpersonal Swimming Institute
P.O. Box 3273
Berkeley, CA 94703

(510) 540-4804
(800) 723-7946 outside California

CARING FOR YOUR GEAR

You just had a great snorkeling experience – now you can thank the gear that helped make it possible, by taking good care of it.

RINSE AND DRY

If there are beach showers, head right up and rinse off. Salt residue is sticky, and corrosive. Rinse it and any sand off your wetsuit, fins, mask, snorkel, and any other gear before the saltwater dries on.

If you can, dry your gear in the shade. It's amazing how much damage sun can do to the more delicate equipment – especially the mask. When he sun odometer hits 100,000 miles, kiss those rubber parts goodbye.

SAFETY INSPECTIONS

Keep an eye on vulnerable parts after a few years (strap, snorkel-holder, buckles). Parts are usually easy to find in Hawai'i, but not in the middle of a snorkeling trip unless you're on a well-equipped boat. We've had to snorkel for an hour with one fin after an aging strap gave way, just as it was stretched too far in the process of gearing up.

If you use any equipment with purge valves, keep an eye on the delicate little flap valves, and replace them when they deteriorate.

Inspect your stretch suit or wetsuit for weakening seams, or the beginnings of splits and tears. You're stretching this gear every time you put it on, and it cannot last forever.

Look out for water droplets on the inside of your watch lens. Once saltwater gets in, battery failure soon follows. Sometimes, baking the watch in the sun to drive out the moisture may turn the trick, and the watch will keep working. If you have a battery replaced, ask if the shop can replace the o-ring gasket, too. This will improve your chances of staying water-tight.

CLEAN YOUR MASK

A mask needs a thorough cleaning between trips as well. Use a regular, non-gel toothpaste to clean the lens inside and out, polishing off accumulated goo. There's no point in going all this way to look at fascinating creatures through little dried-on spots on the lens. Wash the toothpaste off with warm water, using your finger to clean it well.

HAZARDS

Life is not safe. Go for a walk, and meteors fall from the sky and bonk you on the head, occasionally. So, it stands to reason that snorkeling has a few hazards that you should know and avoid if possible.

Obviously you already know the dangers of car and air travel, yet you mustered your courage and decided that a trip to Hawai'i was worth the risks. And you took reasonable precautions – like buckling your seat belt. Well, if you use your noggin, you're probably safer once in the water than you are driving to get *to* the water.

Some people are hesitant to snorkel because they imagine meeting a scary creature in the water. It's actually safer to see them, though, than to be swimming over and around them not knowing what's there.

We don't think it makes sense to overemphasize certain lurid dangers (sharks!) and pay no attention to the more likely hazards of sunburn and stepping on sea urchins (which certainly cause far more aggravation to large numbers of tourists).

Risks are always relative and we often react emotionally rather than rationally at the thought of them. Once on a tranquil Caribbean isle, an old woman warned us to *be careful!* because of all the murders there. She was eager to tell all the details. It turned out that three local people had been killed, by their spouses. She then assured me these heinous crimes had all taken place within just the last 100 years!

SUNBURN

Undoubtedly the worst medical problem you're likely to face – especially if you have the wrong ancestors. Use extra water-resistant sunblock in the water and always wear a shirt during the day. Some people need to avoid the sun entirely from 10:00 to 3:00, but that's a good excuse to go early and avoid the crowds. The top (or open) deck of a boat is a serious hazard to the easily-burned. The best protection is covering up, as evidence mounts that sunscreen still allows skin damage, even though it stops burning.

Sunlight penetrates the water. It also reflects extremely well from water and white sand. In Australia, high skin cancer rates believed due in part to a hole in the ozone layer have heightened awareness of the dangers of reflected sunlight. Kids are being taught to wear light hats with a dark under-brim, as well as dark shirts, to avoid reflecting more rays onto their faces.

It's better not to have sunscreen on your face or hands when putting on your mask, though, because you'll be sorry later when your mask leaks a bit and you get the stuff in your eyes. It can really burn and even make it difficult to see well enough to navigate back to shore.

To avoid using gallons of sunblock, some snorkelers wear lycra body suits. Others simply wear some old clothing, even button-down shirts or old khakis. Loose knits don't work as well because they often balloon out under water – making it tricky to swim.

A few old pro tips: Take an old sun hat to leave on the beach – especially if you have to hike midday across a reflective white beach. Take old sunglasses that are not theft-worthy. If you must leave prescription glasses on the beach, use your old ones. Hawai'i is a great place to find amazingly cheap ($2) sunglasses and flip-flops. For your longer hours in the sun, look into the better sunglasses that carefully filter all the most damaging rays.

RIP CURRENTS

Hawai'i does not have large barrier reefs to intercept incoming waves. Few of the Big Island beaches are well-protected from powerful ocean currents – especially in the winter or during storms.

Waves breaking against a shore push volumes of water up close to the shore. As this piles up, it has to flow back to the ocean, and often flows sideways along the shore until it reaches a convenient, often deeper bottomed exit point. There, a fast, narrow river of water flows out at high speed. Rip currents, which can carry swimmers out quickly, are usually of limited duration, by their very nature, and usually stop no more than 100 yards out.

Sometimes it's possible to swim sideways, but often it's better to simply ride it out. Don't panic. Although the current might be very strong, it won't take you far or drown you, unless you exhaust yourself by swimming against it. It's very easy to float in salt water until help arrives, assuming you're at a beach where someone can see you. Don't try to swim in through waves where there's any chance of being mashed on lava rocks or coral. Don't swim against the current to the point of exhaustion. When in doubt, float and conserve energy.

Even at the most protected beaches (like Kahalu'u), all the water coming in must get out, so there's a current somewhere. Big waves beyond the breakwater may seem harmless, but the more water comes in, the more must get out. This is a good reason to ALWAYS wear fins.

HYPOTHERMIA

Ocean water is always cooler than your body, and it cools you off more rapidly than the air. With normal exertion, your body still cools, bit by bit. After awhile, perhaps 30-45 minutes, you start feeling a little chilly. Later, you begin shivering, and then eventually, hypothermia begins. Your body temperature has dropped so low that your abilities to move and even think begin to get impaired. If your temperature drops too low, you die.

One of the first symptoms of hypothermia is poor judgement, so you need to watch each other – one example of the benefits of having a partner. Check up on each other often in cold conditions. As soon as you are aware that you're cold, it's time to plan your way back. When shivering starts, you should get out of the water ASAP! Be particularly careful in situations requiring all your judgement and skill to be safe – especially when diving, night snorkeling, dealing with waves, or when anticipating a difficult exit from the water.

Usually it's quite easy to warm up rapidly since the air temperature is fairly warm at sea level. If you came by car, it is usually well solar-heated by the time you return, and will help you warm up.

TRUMPETFISH

SEA URCHINS

Probably the most common critter injury is stepping on a spiny sea urchin, and walking away with lots of spines under your skin. The purple-black ones with long spines tend to appear in groups and favor shallow water, so watch carefully if you see even one. Full-foot flippers or booties help a lot, but don't guarantee protection. Watch where you put your hands – especially in shallow water.

While many folks recommend seeing a doctor for urchin spine slivers, others prefer to just let the spines fester and pop out weeks later. Read *The Snows of Kilimanjaro* one more time before trying this self-treatment. You don't want a killer infection setting in.

Remove as much spine as you can, though it's impossible to remove it all. Vinegar (or other acidic liquid) will make it feel better. Periodic soaking in Epsom salts helps, and the small spines will dissolve in a few weeks. See a doctor at any sign of infection. The area will likely turn purple from the dye in the spines, but that will gradually disappear.

Waves are travelling ripples in the water, mostly generated by wind blowing over large expanses of water. Having energy, the waves may travel thousands of miles before dissipating that energy. Here is the wellspring of the beautiful breaking surf. It is also a source of danger.

Take time occasionally to sit on a high point and watch the waves approaching the coast, and you will see patterns emerge. Usually there is an underlying groundswell from one direction, waves that may have originated in distant storms. This is the main source of the rhythmical breaking waves, rising and falling in size in noticeable patterns. Often, there will be a series of small waves, and then one or more larger waves, and the cycle repeats. Pay attention to the patterns, and it will be less likely that you'll get caught by surprise.

Local winds add their own extra energy, in their own directions. In Hawai'i, snorkeling is usually easiest in the mornings, before the daily winds build, and make larger waves in the afternoon.

Occasionally, a set of larger waves, or a single large *rogue* wave comes in with little or no warning. A spot that was protected by an offshore reef suddenly has breaking waves. This is a problem that shouldn't be underestimated.

Our single worst moment in many years of snorkeling and diving was at Brenneke's Beach in Kaua'i after Hurricane Iniki had scattered boulders throughout the beach. We had no problem snorkeling around the boulders in a light swell, protected by reef further out. Suddenly much larger waves crossed the reef and began breaking around us, sweeping us back and forth. Judy was swept into some boulders, leaving bruises on elbows, knees and hip. Fortunately, her wetsuit cushioned the blows enough that no bones were broken.

Since then we have been extra careful to avoid hazardous situations. We always take time to study the waves before entering and ponder what would happen if they suddenly grew much larger, and what our strategy would be. Sometimes we just head for a calmer beach.

BARRACUDA

The Great Barracuda can grow to two meters, has sharp teeth and strong jaws, and swims like a torpedo. For years Judy has removed earrings before swimming after hearing rumors that they attract barracuda, but we've uncovered absolutely no confirming reports of severed ears.

GREAT BARRACUDA

Barracuda can, if motivated, seriously injure a swimmer, and should be taken seriously. Those teeth are just as lethal as they look. Even small barracuda can be aggressive, so don't bother them in any way. They look like they have attitude, and apparently sometimes do. Our own preference is to respect their territory and don't push or pursue them. Others don't worry about them a bit. They're not very common in Hawai'i. Other varieties of barracuda appear more innocuous.

Once a five foot Great Barracuda swam directly beneath our party of four snorkelers in the Caribbean and appeared annoyed that we were invading his home territory (or so we thought from the fierce look on his face). A calm and steady German surgeon headed up the nearest rocks as if she could fly. The rest of us snorkeled by him repeatedly with no problem, but didn't appreciate the look he gave us. We later learned that they always look grumpy, like some folks you may know.

PORTUGUESE MAN-O'WAR

The Portuguese Man-o'War floats on top, looking like a sailfin 1"-4" in size, with long stinging filaments that are quite painful. Stay out of the water if you see one. Even avoid dead ones on the sand! They stay on top of the water with an air-filled bladder and can unroll their long filaments. They're very pretty – shades of purple, but can cause severe pain. Sometimes you don't even see anything – just feel the sting and see long red lines on your body. When Mel encountered one, he felt a whip-like sting on his arm, and a curving red welt appeared.

Vinegar and unseasoned meat tenderizer helps ease the sting and helps stop the release of venom from the stinging cells if tentacles are clinging to you. Use wet sand as a last resort. If you feel very ill, see a doctor right away. Other jellyfish can often be completely harmless. If jellyfish are present, locals will know which ones are a problem. Jellyfish have not been a big problem for us in Hawai'i.

RAYS

Sting rays prefer to avoid you, but hang out on the bottom where they're easy to step on. They prefer resting in calm water that is slightly warmer than the surrounding area – which are just the areas favored by people for swimming. Step on them, and they may sting you, so the injury is usually to the foot or ankle. They can inflict a serious or painful sting to people – especially children. It's best to get immediate first aid, and then medical assistance.

Snorkelers have an advantage in this case over swimmers because snorkelers can see rays and easily avoid them. Swimmers can try shuffling along – giving them a chance to escape. In Maui we've seen them swim between children's legs in shallow water at Kapalua Bay and were amazed to see how adept they were at avoiding people when they are free to do so.

Manta rays don't sting although they're much larger. Of course, they can be bigger than you are, so they certainly can knock you over if you get in their way. Around the Big Island they are often seen six to eight feet across and weigh several hundred pounds., and can get even larger. They maneuver beautifully, so there isn't much danger – unless you're a terrible klutz. Divers though have to be a bit more careful to not have their equipment knocked off inadvertently.

CONE SHELLS

The snails inside these pretty black and brown-decorated shells can fire a poisonous dart. The venom can cause a serious reaction or even death – especially to allergic persons. If in doubt, head for a doctor. This can be easily avoided by not picking up underwater shells in the first place.

POISONOUS FISH

TURKEYFISH
(LIONFISH)

Lionfish (also called turkeyfish) and scorpionfish have spines which are very poisonous. Don't step on or touch them! Their poison can cause serious pain and infection or allergic reaction, so definitely see a doctor if you have a close, personal encounter with one. Better yet, don't! Flippers or booties can help protect your tender feet.

Scorpionfish can blend in so well along the bottom in shallow water that they're easy to miss. Turkeyfish, though, are colorful and easy to spot. Since these fish are not abundant, they are treasured sightings.

SHARKS

Sharks are seldom a problem for snorkelers. In Hawai'i the modest number of verified shark attacks have mostly occurred off O'ahu, with tiger sharks the major perpetrator. As sharks often hunt in very murky river runoff, eating shellfish and whatever else washes down, there is extra risk swimming in times of high runoff. Of course, those are not conditions you're likely to snorkel in, anyway.

Some people will suggest you can pet, feed or even tease certain types of shark. I personally would give sharks a bit of respect and leave them entirely in peace. Most sharks are well-fed on fish, and not all that interested in oiled tourists, but it's hard to tell by looking at a shark whether it has had a bad day. Sharks are territorial, so they can be more aggressive on a first encounter. They especially feed late in the day or at night, causing some people to prefer to enjoy the water more in the morning or midday. If you're in an area frequented by sharks, this might be good to keep in mind.

In Hawai'i divers can see sandbar, black tip and Galapagos sharks. Sandbar sharks will usually swim off quickly, but others might be curious. Get out of the water if they start hanging around or seem curious. In spite of what anyone tells you, most species that are 2' or larger could bite you. Stay perfectly still, or swim steadily away, rather than thrashing around and doing an imitation of a wounded animal.

EELS

Eels are rarely aggressive, and are often tamed by divers. They do possess a formidable array of teeth, which should be avoided. No one would intentionally stick their hand into a hole in the reef to aggravate an eel with those sharp teeth and strong jaws. However, it's easy for the currents and swell to bring you right up close to an eel, surprising you both.

If you are bitten, it's supposedly better to wait till he lets you go, rather than trying to rip yourself away. If you have any personal experience, write us with details. We haven't yet met an actual eel bite victim.

An eel bite can definitely cause serious bleeding requiring prompt medical attention. A good reason not to snorkel alone! Snorkelers see plenty of eels in Hawai'i, and they are easy to find and fascinating. Count on eels to make every effort to avoid you, so there's no need to panic at the sight of one – even if it's swimming freely. Eels aren't interested in humans as food, but they do want to protect themselves and can usually do so with ease by slipping away into the nearest hole.

SNORKELING ALONE

There is one last hazard that you might overlook, in your enthusiasm for the reef. Perhaps your significant other prefers watching sports on ESPN to snorkeling one afternoon, and you're sorely tempted to just head out there alone.

Don't do it. Think about it hard, face the temptation, and just say no.

Snorkeling, done in buddy teams, is a pretty safe recreation, especially if conditions are favorable. Just as in scuba diving, having a buddy along reduces the risk of a small problem becoming a big problem, or even a fatal problem.

Snorkeling done alone increases the risks. We won't spell out all the bad things that could happen; we trust your imagination. If you do go out alone, try to at least limit yourself to well-protected sites under good conditions, and stay near other swimmers and snorkelers.

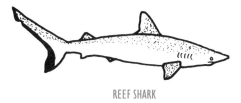

REEF SHARK

I like to watch

"For some reason, the barracuda don't seem scary, any more than the ray does. For some reason, none of this seems scary. Even the idea of maybe encountering a smallish s___k doesn't seem altogether bad.

It's beginning to dawn on me that all the fish and eels and crabs and shrimps and plankton who live and work down here are just too busy to be thinking about me.

I'm a traveller from another dimension, not really a part of their already event-filled world, not programmed one way or another–food or yikes–into their instinct circuits. They have important matters to attend to, and they don't care whether I watch or not. And so I watch."

–Dave Barry

SNORKELING SITES

WHERE ARE THOSE BIG BEAUTIFUL FISH?

The Island of Hawai'i must be well-loved by Madame Pele, for she has made it big. Large! *El Mas Grande*. And she's making it bigger, daily. In the lava department, well endowed indeed. As a person with an inquiring mind, you may want to see it all, and as soon as possible.

If there is a bit of a rolling stone hidden in your normally practical and responsible nature, and your rental car company has given you !!Unlimited Miles!!, you may feel a sudden and uncontrollable urge to circumnavigate the whole volcano, just because it's there. If you give in to this urge, you better get up early. Put the pedal to the metal, and still, you'll probably not make it home before dark. As the miles whiz by, you might ask yourself, 'Am I having fun yet?' and think about getting a life. Mull over some of the other major lifestyle screwups you've made.

The evolved, laid-back snorkelers will already be back home on the Kona Coast relaxing, spinning the morning's fish stories with a cold one at hand, while you're still out there pounding the pavement.

This is because, fortunately for snorkelers and beach lovers alike, the Big Island's ample underwater resources are almost totally concentrated along the Westerly, or 'Kona' Coast. 'Kona' is Hawai'ian for leeward, or protected from the trade winds. Here's where you'll find the calmest and most accessible snorkeling and swimming spots, within an easy drive from condos and "Total Destination" mega-resorts alike.

You can snorkel at an incredible number of spots – especially if you know how to enter the water from the lava rocks carefully and can handle a little (or a lot) of wave action. Even the beaches with sand often have fairly coarse sand or pebbles, so you'll need to venture out with the right gear. Finding easy access and safe conditions is part of the challenge, and we'll help you get in where the colorful action is.

There is an abundance of fish and coral along most of this coast. The scarcity of sandy beaches is actually a plus for snorkeling, because it usually means clearer water. When snorkelers are peering through only 20-30' visibility off the sandy shores of Maui, you may be getting 60-80' sparkling views through the waters of the Big Island. So, enjoy!

In the section ahead, you'll find many snorkeling sites from north to south with more details about our favorites or any with special appeal, such as good beginner beaches.

Many are surprisingly difficult to find, so bring these maps with you. People often drive up and down the highways with no idea what delightful beaches lie just across the barren lava. There is a surprising paucity of signs, so we've included many clues and landmarks to help you find your spot.

Whatever your level of swimming or snorkeling ability, you can find a great spot to enjoy yourself along the sunny Kona Coast. When conditions are right, there are a few other sites farther afield that are worth a try.

RACCOON BUTTERFLYFISH

Passes

In order to comply with Hawai'ian beach access law, the Mauna Kea Beach Hotel (*see page 44*) has a unique pass system allowing a limited amount of beach parking.

Thirty prime Kauna'oa Beach parking spaces are marked 'beach access', and reserved for pass holders. Parking permit 'passes' for these spaces are issued at the guard station as you enter the resort, on a first-come, first-serve basis. Displayed on your dashboard, they allow you to park for the whole day if desired. As you leave, the guard reclaims them, and the space becomes available for another visitor.

On weekends or holidays, **come early** to avoid disappointment. Ten passes are also available for Mau'umae Beach, although as of 1995 you just park alongside a gravel road near the beach access trail. This may change if a new hotel is built in this area.

NOTE: You always have a right to drive down to the head of either trail, drop off or pick up beach goers, and drive back out–one solution if all the passes are in use.

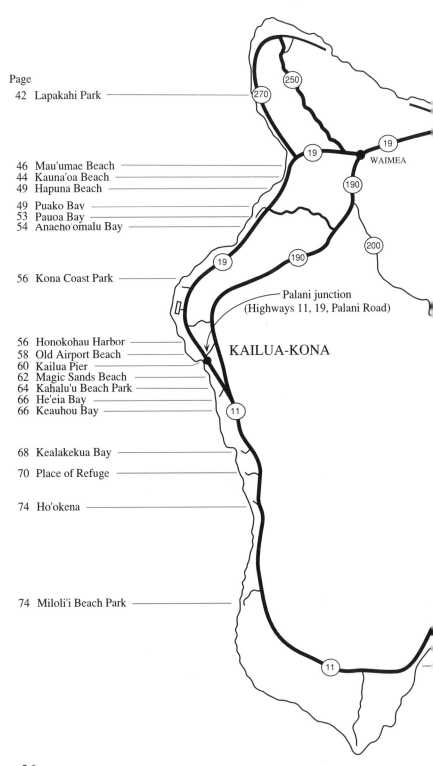

270
250
19
19
WAIMEA
190
19
190
200
19
190

Palani junction
(Highways 11, 19, Palani Road)

KAILUA-KONA

11

11

SNORKELING SITE INDEX

(Also see Road Map on page 4)

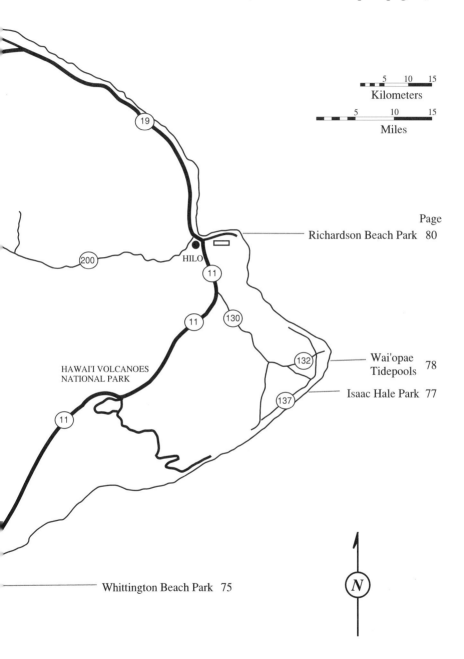

Kilometers
5   10   15

Miles
5   10   15

Page

Richardson Beach Park 80

HILO

HAWAI'I VOLCANOES
NATIONAL PARK

Wai'opae
Tidepools 78

Isaac Hale Park 77

Whittington Beach Park 75

SNORKELING SITES AT A GLANCE

	SNORKELING	ENTRY	SANDY BEACH	RESTROOM	SHOWERS	PICNIC AREA	SCENIC
Lapakahi Park	A	3		•		•	A
Mau'umae Beach	A	1	•			•	A
Kauna'oa Beach	A	1	•	•	•	•	A
Hapuna Beach	B	2	•	•	•	•	A
Puako Bay	A	3					A
Pauoa Bay	A	1	•	•	•	•	A
Anaeho'omalu Bay	C	1	•	•	•	•	B
Kona Coast Park	B	1	•	•		•	A
Old Airport Beach	B	2	•			•	B
Kailua Pier	C	1	•	•	•	•	C
Magic Sands Beach	C	2	•			•	B
Kahalu'u Beach Park	A	1	•	•	•	•	A
He'eia Bay	C	2					C
Keauhou Bay	C	1	•	•	•	•	B
Kealakekua Bay	A	1-3				•	A
Place of Refuge	A	1-3	•	•	•	•	A
Ho'okena	B	3	•	•	•	•	C
Miloli'i Beach Park	A	3	•	•	•	•	B
Whittington Park	C	3	•	•	•	•	B
Isaac Hale Park	C	1	•	•	•	•	B
Wai'opae Tidepools	A	1					A
Richardson Beach Park	A	1	•	•	•	•	A

KEY:

1	Easy	A	Excellent
2	Moderate	B	Good
3	Difficult	C	Fair

LIFEGUARD	SHADE	PAGE	MAP PAGE	NOTES
		42	43	Only when calm; advanced. Historical.
	•	46	47	Secluded, gorgeous. Take our map!
	•	44	45	Beautiful in every way, serene.
•	•	49	48	Afternoon waves; popular with locals.
	•	49	48	Very wide, shallow reef is hard to cross.
		53	52	Requires long, rocky hike around hotel.
	•	54	55	Easy, all amenities, big bay, boating.
•	•	56	57	Spacious, some waves, often uncrowded.
	•	58	59	Best when calm, watch out for sea urchins!
	•	60	61	Quite easy and calm, but near large boats.
		62	63	Sand disappears in winter. Good when calm.
•	•	64	65	Like a swim-in aquarium. Calm, don't miss!
	•	66	67	Small, better snorkeling than swimming.
	•	66	67	Harbor area, good spot for picnic.
		68	69	Easiest from boat. Very calm and protected.
	•	70	71	Several beautiful and varied sites. Don't miss!
	•	74	69	Facilities are not appealing.
	•	74	36	Access very difficult by car. Best by boat.
	•	75	37	Windy, currents. Experts only.
	•	77	76	Fine when calm; excellent park.
		78	79	Large group of interconnecting tidepools.
•	•	80	81	Somewhat shallow (5-10 feet), protected.

As long as you've driven this far north, consider continuing to the end of the highway to Pololu Lookout at the second (yes, second) mile marker 28. This is an excellent road with practically no traffic. It's very beautiful and lush, with a spectacular view of the sharp cliffs at the end. This stretch offers a chance to see some of the differing climate zones that make the Big Island so fascinating. There are thirteen recognized climate zones, and Hawai'i has eleven of them!

There's a house at the end and when we drove by, someone had set up an honor system lemonade stand on the side of the road. Several little towns along this road offer excellent places to stop for lunch or even a bit of crafts shopping. Linger awhile and enjoy this quiet, friendly and beautiful corner of the Big Island.

PETROGLYPH OF PADDLERS

Fish Psychology 101

"Many fish are swimming right up and giving me dopey fish looks, which basically translate to the following statement: "Food?" That's what fish do all the time–they swim around going: "Food?" You can almost see the little questions marks over their heads. The only other thought they seem capable of is: "Yikes!"

Fish are not known for their SAT scores. This may be why they tend to do their thinking in large groups. You'll see a squadron of them coming toward you, their molecule-size brains working away on the problem ("Food?" "Food?" "Food?" "Food?"); and then you suddenly move your arm, triggering a Nuclear Fish Reaction ("Yikes!" "Yikes!" "Yikes!" "Yikes!") and FWOOOSSHH they're outta there, trailing a stream of exclamation marks."

–Dave Barry

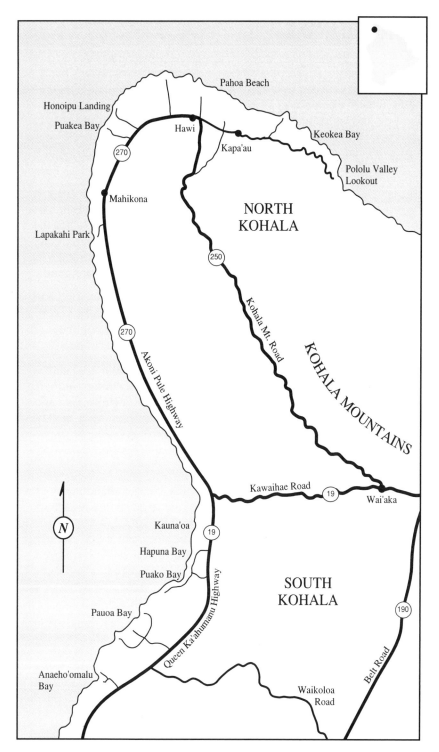

Honoipu Landing

Puakea Bay

270

Hawi

Pahoa Beach

Kapa'au

Keokea Bay

Pololu Valley
Lookout

NORTH
KOHALA

Mahikona

Lapakahi Park

250

Kohala Mt. Road

KOHALA MOUNTAINS

270

Akoni Pule Highway

N

Kawaihae Road

19

Wai'aka

Kauna'oa

19

Hapuna Bay

Puako Bay

SOUTH
KOHALA

Queen Ka'ahumanu Highway

190

Pauoa Bay

Belt Road

Anaeho'omalu
Bay

Waikoloa
Road

LAPAKAHI PARK

Lapakahi Marine Life Conservation and State Historical Park has a lovely, serene beauty. Way up in North Kohala, it can seem abandoned in the winter when waves tend to be big. There is no running water or showers and little shade on the trail, but there is a cooler of water in the main exhibit area. The counter with literature was manned by just a cat when we were there, but he seemed to be taking his role quite seriously. Portable toilets are available.

It's about a 250 yard hike directly down to the water – worth the hike even if you don't swim. The trail is on the side of moderate hill. The trail isn't well marked, but the lack of trees means it's easy to see the beach. As you near the shore and intersect the main trail parallel to the shore, turn left and enter at the small pebble "beach" Snorkel whichever section appears calmest, but only if conditions are good.

Lapakahi has excellent snorkeling, but access is across the coral and rocks, so don't try if there's any problem with swell. It can be very tricky getting back out if big waves pick up. Mornings would be a better bet here, since the wind often picks up in the afternoon.

Good swimming skills and fins are definitely essential here. Don't venture too far from shore since the currents can be strong out there. This is a place where you need to sit and watch for awhile. Also ask local people for advice if possible, keeping in mind that they may be much better swimmers than you are. Wave size often varies in noticeable cycles, but may also change suddenly (*see Understanding waves, page 29*).

Lapakahi is the site of a 13th century fishing village, abandoned in the 1800's when water was diverted for use at the sugar plantations. Stone foundation are all around you and it's now a hot, dry, but very beautiful site. It's hard to compare lovely beaches, but this is definitely one of the most beautiful – sit and imagine what it might have been like to live and work in this charmed spot.

GETTING THERE *Go north on Highway 19 (see map, page 41), past mile marker 68, and then turn left on Highway 270 near the big harbor. You're at mile marker 2 on this road. Watch immediately for a Y (at a gas station) and continue on 270 (the right of the Y) heading north. Follow 270 north to the Lapakahi sign on the left. It's between mile markers 13 and 14. Notice that they lock the gate at 4 p.m. and the sign suggests that*

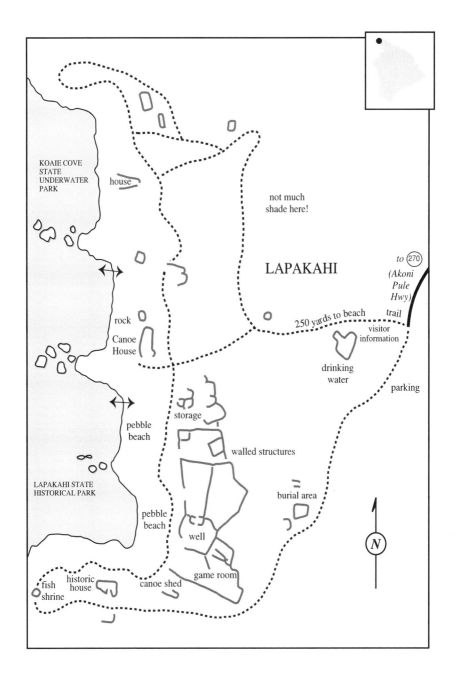

KOAIE COVE
STATE
UNDERWATER
PARK

house

not much
shade here!

LAPAKAHI

to (270)
(Akoni
Pule
Hwy)

rock

Canoe
House

250 yards to beach   trail

visitor
information

drinking
water

parking

storage

pebble
beach

walled structures

LAPAKAHI STATE
HISTORICAL PARK

burial area

pebble
beach

well

N

fish
shrine

historic
house

canoe shed

game room

*if locked in, you may have to hike a minimum of 7 1/2 miles north to the
nearest phone! The total distance from the Kailua-Kona Palani junction
(see map, page 61) to the Lapakahi turnoff is 46.6 miles.*

KAUNA'OA BEACH

In early 1960s, the vast barren sharp a'a lava fields of Kohala seemed an unlikely and inhospitable spot for hotels. Though the area is fringed with beautiful beaches, they were hard to get to unless approached by boat. They lay quiet in the sun, with occasional visits by local people who knew of their beauty.

The solitude disappeared forever in 1965, when the first luxury hotel complex along the Kohala Coast was built on beautiful Kauna'oa Beach by L. Rockefeller: the Mauna Kea Beach Resort. This oasis of luxury among the lava fields drew a rich and famous clientele for many years, and has spawned a host of other posh resorts, each seemingly larger than the last. The Mauna Kea itself underwent a thorough remodeling in the early 1990s. The beach chosen for this pioneer remains one of the best.

Kauna'oa Beach has excellent snorkeling as well as sunning, and is an unforgettable day trip. A beautiful long crescent of soft white coral sand fringed with palms is set off by dark turquoise water. There is plenty of room for hotel guests and visitors alike. The waters are usually safe and calm, but pay attention to hotel signs about current conditions.

Snorkel either side out to the point – whichever appears calmest. The left has more coral, but the fish are great on either side. We saw turtles, eels, and a huge variety of fish.

This is a very protected bay with easy access from the sand. It gets a bit choppy out at the points, but you don't have to swim that far. The excitement starts almost as soon as you enter the water.

LONGNOSE BUTTERFLYFISH

GETTING THERE *Go north on Highway 19* (see area map, page 48), *past several major resorts, and on past the turnoff for the town of Waikoloa. Continue past Hapuna Beach, to just past mile marker 69 (32 .6 miles north of the Kailua-Kona Palani junction). At the resort entrance,you'll see a guard station. Ask for a beach "pass" (*see **Passes**, page 35). *With pass in hand, drive down the hill, curving left, on past the hotel and large parking lot, and enter the last parking lot at road's end. Drive on to the end of the lot, where you'll see marked beach access parking spaces. A wide paved path to the beach begins there. On your left, you'll see the first-rate showers and restrooms meticulously maintained by the Mauna Kea.*

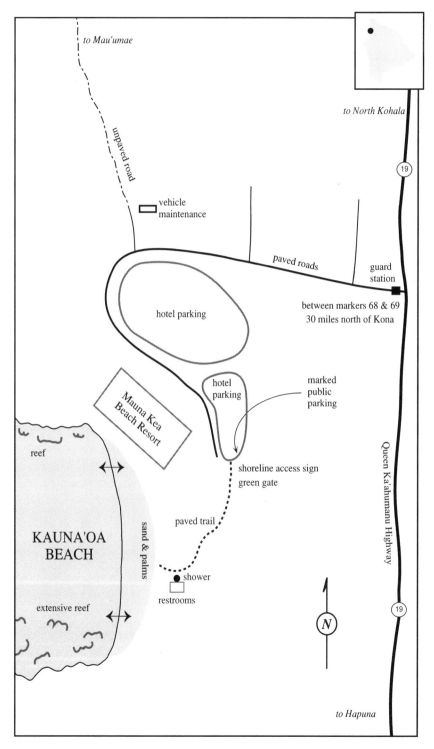

to Mau'umae

unpaved road

vehicle
maintenance

paved roads

to North Kohala

19

guard
station

between markers 68 & 69
30 miles north of Kona

hotel parking

hotel
parking

marked
public
parking

Mauna Kea
Beach Resort

reef

shoreline access sign
green gate

paved trail

KAUNA'OA
BEACH

sand & palms

extensive reef

shower

restrooms

Queen Ka'ahumanu Highway

19

N

to Hapuna

45

MAU'UMAE BEACH

If the 30 passes for the Mauna Kea are in use, consider asking the same guard station (*see Kauna'oa Beach, page 44*) for one of the ten passes to Mau'umae Beach.

Mau'umae Beach is similar to Kauna'oa, but smaller and undeveloped. With no hotel and no facilities, it also has few people, but plenty of fish and coral. (There is talk of a hotel in planning, so access may change – if it has, ask for directions at the guard station).

This is a gorgeous, secluded beach with terrific snorkeling and swimming. Just snorkel along the rocks on the left as far as the point. It's calm, clear and you might even have the whole beach to yourselves if you arrive a bit early (bathing suits seem optional). No restrooms or showers, but no crowds either. A real gem. Anyway, you can always take the short drive down to the Mauna Kea path to find these amenities after your swim.

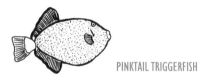

PINKTAIL TRIGGERFISH

GETTING THERE   *From the guard station* (see area map, page 48)*, drive down the hill 1/2 mile, and take the third right turn (just before the road makes a sweeping 90° turn to the left). The third paved road gives the appearance of a hotel service entrance, and has a sign saying "Private Road". Not to worry! In .15 mile along this road, you'll see buildings and lots of service vehicles on your right just before the paving ends.*

Continue on, holding to the left through a chain link fence, onto the unpaved road. Now you are out in the lava field countryside. The road is gravel, but not difficult. Cross two small wood plank bridges and pass a private road on the left.

At exactly 1/4 mile from where the gravel road started, park along the road, remembering to display your "beach pass" on the dashboard. A small path on the left, near phone pole #22, leads through the shrubs toward the ocean. After about 125 yards, you'll come to a marker saying "Ala Kahaka", where you must turn left or right.

Take the left turn for another 125 yards through some small overhanging trees and you're there. (If you turn right at the T, you'll eventually come to a small rocky cove where a creek comes down–not the best snorkeling access or prospects, but a possibility if you want a private spot to sun).

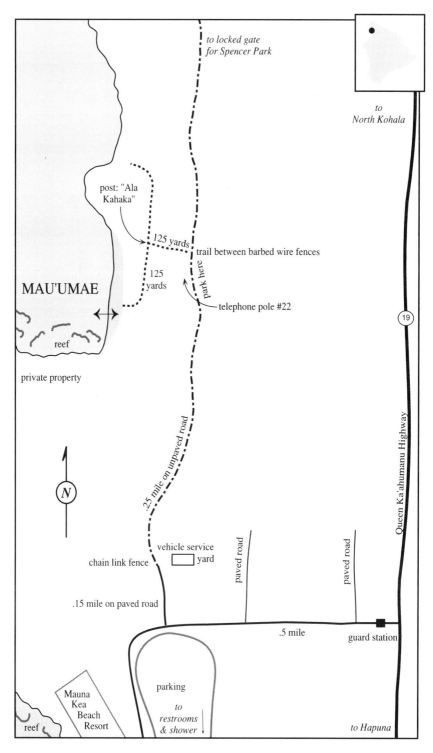

to locked gate
for Spencer Park

to
North Kohala

post: "Ala Kahaka"

125 yards

125 yards

trail between barbed wire fences

park here

telephone pole #22

MAU'UMAE

reef

private property

.25 mile on unpaved road

N

chain link fence

vehicle service yard

.15 mile on paved road

paved road

paved road

Queen Ka'ahumanu Highway

19

.5 mile

guard station

Mauna Kea Beach Resort

parking

to restrooms & shower

reef

to Hapuna

47

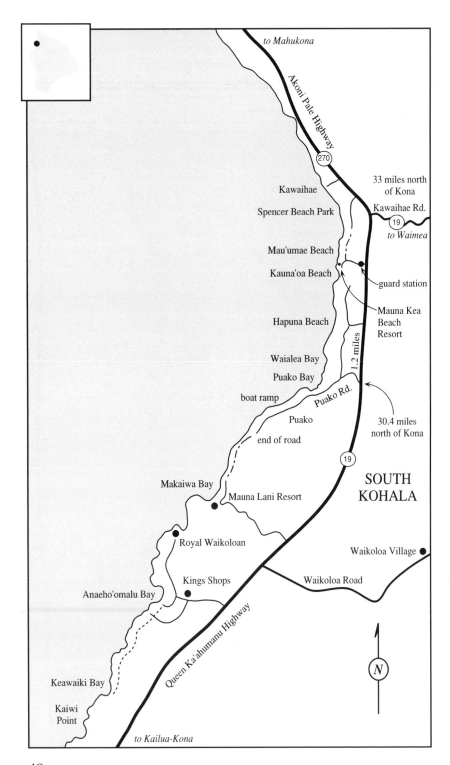

to Mahukona

Akoni Pule Highway

270

33 miles north
of Kona

Kawaihae

Spencer Beach Park

Kawaihae Rd.

19

to Waimea

Mau'umae Beach

Kauna'oa Beach

guard station

Mauna Kea
Beach
Resort

Hapuna Beach

1.2 miles

Waialea Bay

Puako Bay

boat ramp

Puako Rd.

Puako

30.4 miles
north of Kona

end of road

19

SOUTH
KOHALA

Makaiwa Bay

Mauna Lani Resort

Royal Waikoloan

Waikoloa Village

Kings Shops

Anaeho'omalu Bay

Waikoloa Road

Queen Ka'ahumanu Highway

Keawaiki Bay

Kaiwi
Point

N

to Kailua-Kona

48

HAPUNA BEACH

Hapuna is a large public beach, with restrooms, and plenty of parking and picnic tables. This is one of the Big Island's longest white sand beaches and is very popular with local folks for picnics and fun. It is also reputed to be one of the most dangerous when the surf is up.

The surf can be calm as glass in the morning, then pick up quite suddenly as the day wears on. For snorkelers who can't be watching the waves and fish at the same time, this can sneak up on you. During calm conditions, there is snorkeling at both ends. Choose according to wave and wind direction.

Hapuna is certainly a delightful place to stop for a picnic or a post-swim shower. There is also a 3/4 mile long scenic old Hawai'ian trail going north from Hapuna, and you may choose to enjoy the scenery and lively social scene here, rather than gamble with heavy waves.

GETTING THERE   *Go north on Highway 19 past several major resorts, and on past the turnoff for the town of Waikoloa. Soon you'll see a big sign for "Hapuna Beach.", 31.6 miles north of the Kailua-Kona Palani Road junction.*

PUAKO BAY

Puako Bay, stretching for several miles south of Kauna'oa Beach, has a true fringing reef, the only one as yet developed on the Big Island. The broad, quite shallow reef extends several hundred yards out into the bay. It protects the immediate shoreline, but can pose a safety problem for snorkelers, other than on that rare, flat as a mirror day.

To get to the fine snorkeling beyond the reef, you have to swim out over the shallow shelf. If the surf is up (or comes up while you're out), you risk having large waves beat you on the reef. This is one of the ways sand is made, and its a good opportunity to become a personal part of the scenery. (Leaving pieces of your bones is not considered to be littering).

Rogue sets of waves are common here (especially in the winter), so we'd recommend a lot of caution (*see Understanding Waves, page 29*). It may be possible to swim out at the boat ramp noted on the map, or from the south end of the road, if conditions are good. Unless you are one tough, experienced snorkeler, this is one of the spots better approached from a boat.

The deeper area just beyond the reef offers a chance to see large pelagic fish, as well as other larger deep water creatures, including sharks. Some snorkelers and surfers like this, while others feel a little too close to the food chain for comfort.

GETTING THERE   *From Highway 19 (see map, page 48), take Puako Road toward the ocean, 30.4 miles north of the Kailua-Kona Palani junction. It has a sign, but is easily missed. Continue generally south, ignoring the beach signs urging you to turn right (which would take you to a different beach).*

A long, almost solid row of houses has been built between the road and the shoreline. Every so often you'll see a "beach access" sign between two houses, with a short path to the water. Take your choice. The trick is to park without blocking anyone's driveway.

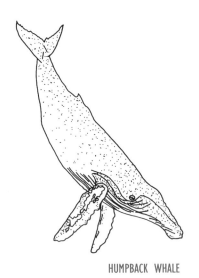

HUMPBACK WHALE

If you are swimming along snorkeling peacefully, and your vision suddenly gets totally out of focus, don't be too quick to panic and call for a doctor. While you may have had a stroke, there is a much more likely cause:

You've probably just entered into an outdoor demonstration of the refractive qualities of mixtures of clear liquids of different densities. Is that perfectly clear?

Near the edge of some protected bays, clear spring water oozes smoothly out into the saltwater. As it is lighter than the mineral-laden saltwater, it tends to float in a layer near the surface for a time.

Now, clear spring water is easy to see through, as is clear saltwater. If you mix them thoroughly, you have dilute saltwater, still clear. But when the two float side by side, the light going through them is bent and re-bent as it passes between them, and this blurs your vision. It's much like the blurring produced when hot, lighter air rises off black pavement, and produces wavy vision and mirage.

These lenses of clear water drift about, and often disappear as quickly as they appeared. Swimming away from the source of the spring water usually solves the problem. Clear at last?

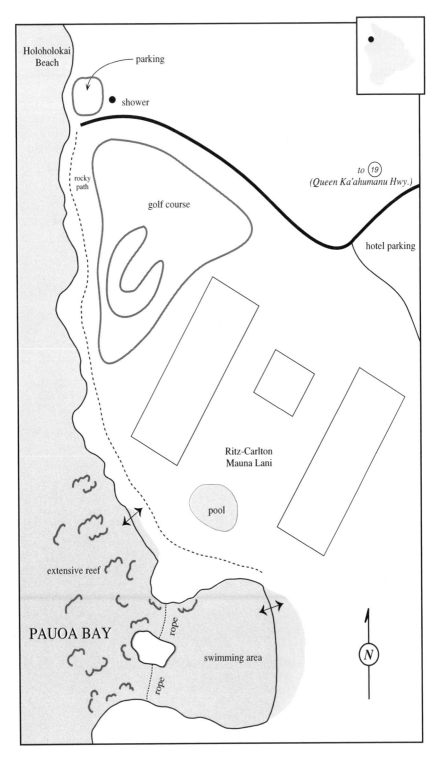

Holoholokai
Beach

parking

● shower

rocky
path

golf course

to ⑲
(Queen Ka'ahumanu Hwy.)

hotel parking

Ritz-Carlton
Mauna Lani

pool

extensive reef

PAUOA BAY

rope

rope

swimming area

N

PAUOA BAY

The Ritz-Carlton Mauna Lani Resort has created an elegant tropical oasis surrounding the small, pretty white sand beach of Pauoa Bay.

This is a good place to snorkel, quite protected and very interesting. The rocky areas outside the bay are better for snorkeling than swimming, because of sharp rocks, coral and sea urchins in shallow water areas. We've seen quite a bit of fresh water runoff creating a foggy (some think it's oily) look to the water in places (*see Doctor my eyes, page 51*).

Still, there's lots to see such as Picasso or lagoon triggerfish – our cover fish, and one of our favorites in Hawai'i. The small bay is well-protected and not overly deep, lending itself to easy snorkeling.

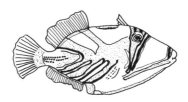

PICASSO TRIGGERFISH

GETTING THERE   *The turnoff for the Mauna Lani Resort is easy to find between mile 73 and 74 on Highway 19 (*see area map, page 48*). However, entry to the beach is another matter. It would be nice if the Resort provided beach access paths or passes; in their absence, access is difficult. We parked at Holoholokai Beach County Park, which is the next beach north along the coast road from the hotel. It's another pretty spot for a picnic, with showers and restrooms.*

To reach Pauoa Bay, take a 1/4 mile hike south on a rough lava path along the water. You'll quickly see why shoes are required – not just flip-flops. This is sharp, rough, loose rock! Follow the coast south, watching out for stray golf balls and in places golf carts. As you near the hotel, choose a spot to enter the water. It's actually easier to enter the water at the alternative entry point shown on the map on the opposite page, across from the pool, rather than in the roped off swimming section where sea urchins are more of a problem.

ANAEHO'OMALU BAY

The public park at Anaeho'omalu Bay is pretty, calm and very popular. Entry is very easy all along this white sand beach and you can easily snorkel around scattered boulders and some coral in 5-10' deep water. Because of the large number of people, the coral has taken a beating here. The visibility can be limited, due to the sandy bottom getting stirred up by the winds that blow regularly along the North Kohala coast. The crowds may put some folks off.

Still, most people are taken by the wide and pretty beach with plenty of palms for shade. It's certainly a good place for children and beginners. There are enough fish to make it interesting for everyone. Snorkel just about anywhere among the coral heads.

The parking area is bare gravel, but you walk out into a lovely park with nice showers, restrooms, landscaping and a huge bay full of boats, people, sports of all kinds, as well as coral and pretty fish.

GETTING THERE   *At the 76 mile marker on Highway 19 (*see area map, page 48*) take the big turnoff toward the huge resorts (including the Royal Waikoloan). This intersection is 24.5 miles from the Kailua-Kona Palani junction. When you see the King's Shops on your right, take a sharp left just across from the shops.*

Continue on, through an open gate, and drive until you reach the very last parking area right near the water.

CONVICT TANG

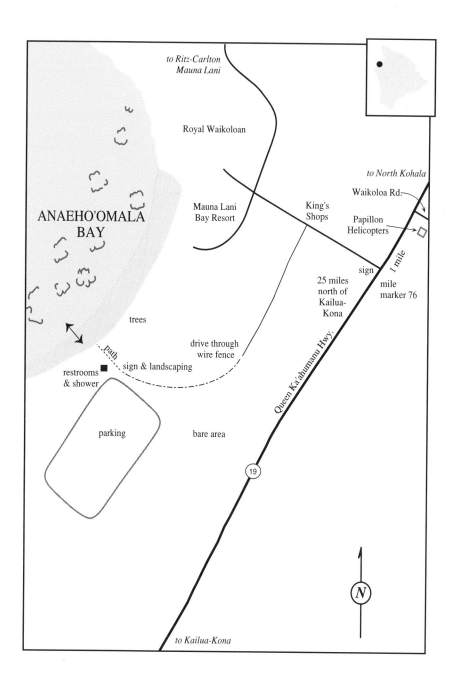

to Ritz-Carlton
Mauna Lani

Royal Waikoloan

to North Kohala

Waikoloa Rd.

ANAEHO'OMALA
BAY

Mauna Lani
Bay Resort

King's
Shops

Papillon
Helicopters

sign

1 mile

25 miles
north of
Kailua-
Kona

mile
marker 76

trees

path

drive through
wire fence

sign & landscaping

restrooms
& shower

Queen Ka'ahumanu Hwy.

parking

bare area

19

N

to Kailua-Kona

KONA COAST PARK

Formerly called Mahai'ula. The access road takes you directly to the beach, which is large, pretty and still fairly undeveloped. Sometimes the waves are quite large in the winter, however, the reefs are large and offer protection when surf isn't too high.

Walk to the right near the lifeguard station, then snorkel almost anywhere to the right of this. It's a big area and we had the whole place to ourselves. We particularly enjoyed the sea turtles. The water was a bit choppier here and a bit murky, but still very interesting with plenty of good-sized fish.

For a very pretty and even more isolated beach, continue walking north to Makalawena Beach. It's more than a quarter mile, but all white sand along the way. You'll see several little coves, all connected by a sand beach. This is also a beautiful picnic spot. There are no facilities, but lots of lovely sandy beach, and good snorkeling. This whole rocky Kohala Coast has reasonable snorkeling almost everywhere. It's mostly a matter of finding easy access and avoiding waves and currents.

GETTING THERE   *Take Highway 19 north exactly 2.7 miles past the Keahole Airport entrance. Watch carefully for a small road to the left, 9.7 miles from the Kailua-Kona Palani junction. There is a sign, but it's easy to miss. There's a 1 1/2 mile long rough road built on some rugged lava heading toward the ocean – actually an interesting approach. Don't show up on a Wednesday or before 9:00 a.m., because the gate will be locked and it's a very long hot hike to get in.*

SADDLEBACK BUTTERFLYFISH

HONOKOHAU HARBOR

Harbor snorkeling is, well, harbor snorkeling. For those who want to have a chance of seeing sharks, it might be worth a try. We prefer not to snorkel with so many boats and fumes and junk in the water.

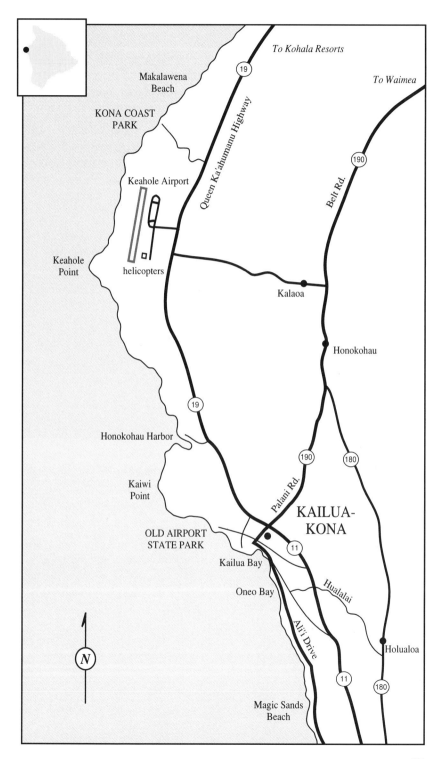

To Kohala Resorts

To Waimea

Makalawena
Beach

KONA COAST
PARK

19

Queen Kaʻahumanu Highway

190

Belt Rd.

Keahole Airport

helicopters

Keahole
Point

Kalaoa

Honokohau

19

Honokohau Harbor

Kaiwi
Point

190

180

Palani Rd.

KAILUA-
KONA

OLD AIRPORT
STATE PARK

11

Kailua Bay

Oneo Bay

Hualalai

Aliʻi Drive

Holualoa

N

11

180

Magic Sands
Beach

OLD AIRPORT BEACH

We snorkeled this site in very small swells, but found the entry quite difficult because the rocks were slippery and there were many spiny sea urchins to fall on. In fact, we helped another tourist remove spines from her foot and retrieved her friend's flip-flop from the sea.

We did see plenty of fish, but were not impressed with the coral. In winter I wouldn't recommend this for beginners. In calm water, you can continue to snorkel to the right around the point and along a rocky ledge. It's very beautiful. Lots of snorkel and dive boats come here regularly.

This is a nice picnic spot and a local hangout, which lends color, as long as the guys hanging out drinking beer don't make you nervous. And if you always wanted to drive on an airport runway, here's your chance!

PARROTFISH

GETTING THERE   *Take the Kuakini Highway north from Palani Road (the main drag in Kailua-Kona). This intersection is more or less the center of town. Follow Kuakini north.*

You'll pass playing fields on your left, then the road eventually becomes an airport landing strip. Drive on the landing strip to the far end (at the north). Park here and walk out a short way to the end of the beach. It isn't far, but wear shoes.

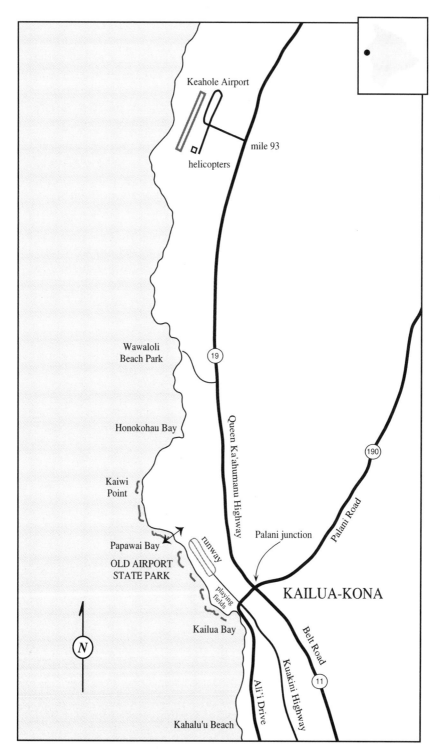

Keahole Airport

mile 93

helicopters

Wawaloli
Beach Park

Honokohau Bay

Kaiwi
Point

Papawai Bay

OLD AIRPORT
STATE PARK

Kailua Bay

Kahalu'u Beach

19

Queen Ka'ahumanu Highway

190

Palani Road

Palani junction

runway

playing
fields

KAILUA-KONA

Ali'i Drive

Kuakini Highway

Belt Road

11

N

KAILUA PIER

Kailua-Kona is where the action is on the Kona Coast – you'll go there to shop, to eat, or to hang out. This bustling town has grown up, or perhaps sprawled around a core of historical buildings that deserve a visit during those odd hours that you're not snorkeling.

The Kailua harbor offers great protection and extremely easy entrance as long as you don't mind sharing the water with large boats. There's actually good snorkeling if you follow the pier, then continue up the coast to the north. The easiest entry is via the King Kamehameha Hotel beach, just west of the pier. If small waves are breaking onto the coast, it's still easy to stay beyond them.

The entire harbor area is almost always calm – something to remember on a stormy day. Use fins anyway because you might want to swim quite a long ways. (This happens to be the site of the Ironman Triathlon competition, so there's no chance to snorkel when this popular event takes place).

GETTING THERE   *Take Highway 19 south past the airport, till you see the signs at Palani Road for Kailua-Kona. Turn west at the Palani junction to drop down into town. As you pass the imposing King Kamehameha Hotel on the right, and the road takes a 90° turn left, the pier is straight ahead. At this corner, Palani Road becomes Ali' i Drive.*

Parking in town can look impossible, but if you keep trying, you can usually find space in one of the lots up the hill from Ali' i Drive.

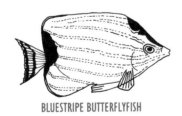

BLUESTRIPE BUTTERFLYFISH

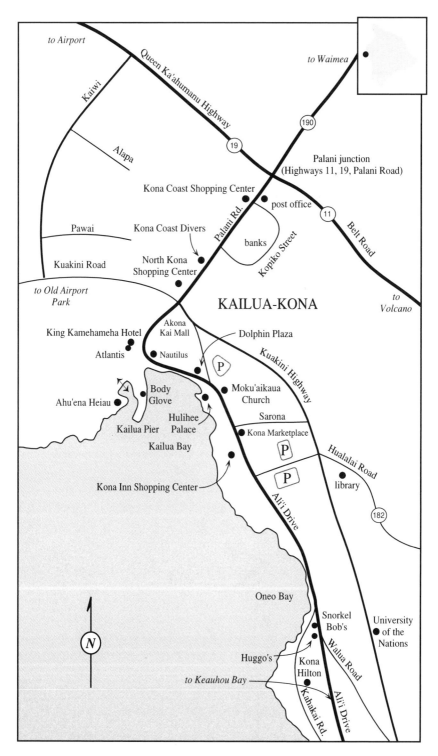

to Airport

Queen Ka'ahumanu Highway

to Waimea

190

Kaiwi

Alapa

19

Palani junction
(Highways 11, 19, Palani Road)

Kona Coast Shopping Center

post office

11

Belt Road

Pawai

Kona Coast Divers

Palani Rd.

banks

Kopiko Street

Kuakini Road

North Kona
Shopping Center

to Old Airport
Park

to
Volcano

KAILUA-KONA

King Kamehameha Hotel

Akona
Kai Mall

Dolphin Plaza

Kuakini Highway

Atlantis

Nautilus

P

Ahu'ena Heiau

Body
Glove

Moku'aikaua
Church

Kailua Pier

Hulihee
Palace

Sarona

Kailua Bay

Kona Marketplace

P

Hualalai Road

P

library

Kona Inn Shopping Center

Ali'i Drive

182

Oneo Bay

Snorkel
Bob's

University
of the
Nations

N

Huggo's

Kona
Hilton

Walua Road

to Keauhou Bay

Kahakai Rd.

Ali'i Drive

MAGIC SANDS BEACH

Magic Sands Beach (also known as Disappearing Sands or White Sands Beach) is easy to find, and handy. It's a short walk from many condos and really the only safe choice here for swimming or snorkeling if you don't drive. People do snorkel and sometimes swim straight off the lava rocks, but that can lead to real trouble if the surf picks up. So leave such risky business to the most experienced local swimmers, and try White Sands as your entry point instead.

As these various names indicate, sand can vanish quickly in a winter storm. It can all disappear overnight, just like magic.

When ample sand is in place, it's a pretty little beach, popular for swimming and body-surfing – especially for children. When the waves aren't too high, it's easy to swim beyond them and snorkel this area toward the south along the rocks or north around the point. You don't have to go far to avoid the crowds. We saw very few snorkelers here – often just a couple.

GETTING THERE   *Take Ali'i Drive south from Kailua-Kona about four miles from the King Kamehameha Hotel, where Palani Road becomes Ali'i Drive. You'll pass lots of condos and this is the first real beach you'll see. It's right along the road.*

ORANGEBAND SURGEONFISH

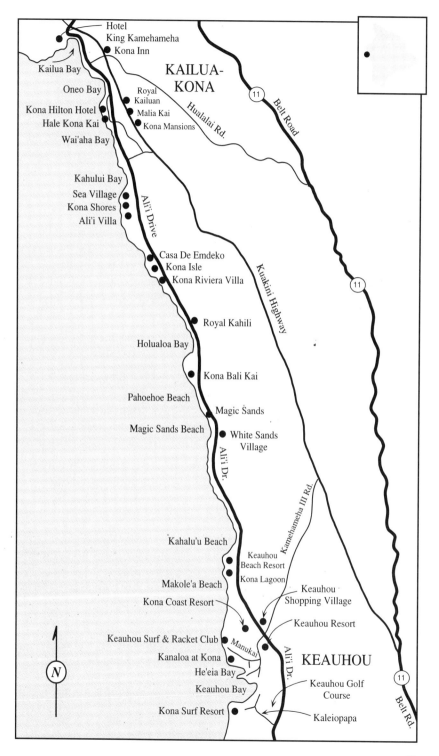

Hotel
King Kamehameha
Kona Inn

KAILUA-
KONA

Kailua Bay

Oneo Bay

Royal
Kailuan
Malia Kai
Kona Mansions

Hualalai Rd.

11

Belt Road

Kona Hilton Hotel
Hale Kona Kai
Wai'aha Bay

Kahului Bay
Sea Village
Kona Shores
Ali'i Villa

Ali'i Drive

Casa De Emdeko
Kona Isle
Kona Riviera Villa

11

Kuakini Highway

Royal Kahili

Holualoa Bay

Kona Bali Kai

Pahoehoe Beach

Magic Sands

Magic Sands Beach

White Sands
Village

Ali'i Dr.

Kahalu'u Beach

Kamehameha III Rd.

Keauhou
Beach Resort

Kona Lagoon

Makole'a Beach

Kona Coast Resort

Keauhou
Shopping Village

Keauhou Resort

Keauhou Surf & Racket Club

Kanaloa at Kona

Manukai

He'eia Bay

Keauhou Bay

Kona Surf Resort

Ali'i Dr.

KEAUHOU

11

Belt Rd.

Keauhou Golf
Course

Kaleiopapa

N

KAHALU'U BEACH PARK

Kahalu'u Beach, sometimes called "Children's Beach", is adjacent to the road, and clearly marked. It has plenty of parking, white sand, showers, restrooms, a covered picnic area, lifeguard and a nice curve of protecting reef, creating a most unusual walk-in aquarium.

This is a perfect site for beginners because it's calm, has easy entry if you don't mind snorkeling around lots of legs, and an abundance of unusually large, gorgeous fish. It's all fairly shallow and the fish are often fed, so they are tame and can be seen up-close. This makes for excellent photography as well.

Turtles also hang out here and seem able to ignore the tourists. The shallowness makes beginners more comfortable. Come early or late, avoid weekends, or swim beyond the crowds and this is a small snorkeler's heaven. Just get off the plane, grab swimsuit, mask, snorkel and fins (even if it's late afternoon) and head for Kahalu'u. It's the quickest way to start enjoying your vacation.

If you're on the Big Island and have never tried snorkeling, this is a perfect chance to give it a try. Even for the pros, don't miss this one!

Some people may be put off by the crowds, the fed fish, or the flattened reef tops from too much reef walking. This is not a wilderness experience. Despite that, if you pick your time right, and you swim a little ways out, the superb collection of large specimens of spectacular fish still makes this spot worth a visit.

As well as turtles, we have seen parrotfish, Picasso triggerfish, groups of scrawled filefish, Moorish idols, raccoon butterflyfish, Achilles tang, threadfin butterflyfish, blue-stripe snapper, ornate butterflyfish, pinktail triggerfish, yellow tang, trumpetfish, spotted trunkfish – well, you get the idea.

A word of caution though: when the surf kicks up, even if it's stopped by the breakwater, there's still plenty of current caused by the water coming in. The current has a tendency to sweep north and then out, so stay closer to shore during storms and wear your fins.

If caught by the current, don't panic. It's much better to float and wait for help than to try to swim back across rocks or lava. At Kahalu'u Beach Park, at least, someone is likely to see you. Keep in mind that rip currents usually don't take you more than a few hundred feet (*see Rip currents, page 27 for more details*). It's just a matter of all that water pouring in and needing somewhere to go out.

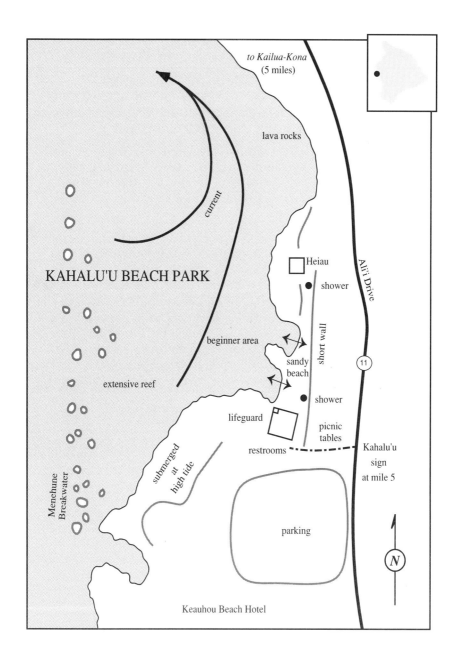

to Kailua-Kona
(5 miles)

lava rocks

current

KAHALU'U BEACH PARK

Heiau

shower

beginner area

short wall

sandy
beach

extensive reef

shower

lifeguard

picnic
tables

restrooms

Kahalu'u
sign
at mile 5

submerged
at
high tide

Menehune
Breakwater

parking

N

Keauhou Beach Hotel

Ali'i Drive

11

GETTING THERE *Take Ali' i Drive south from Kailua-Kona to the 5 mile mark (about a mile south of Magic Sands Beach) and you can't miss this small, but outstanding snorkeling beach.*

KEAUHOU BAY

The Fair Winds and Sea Paradise offices are located at the edge of Keauhou Bay. At the Keauhou Pier you'll find the Fair Wind office on the left, and the smaller Sea Paradise office on the right.

If you don't mind some boat traffic (although it's not frequent), this is an excellent place to snorkel and swim. There's a small park with a tiny beach, so enter here. Choose whichever side is calmest. There's plenty of reef here, but stay inside the harbor away from the currents and surf.

Winter can sometimes bring big waves crashing spectacularly across the bay against the point just below the Kona Surf Resort.

GETTING THERE   *From the corner of Ali' i Drive and Kamehameha III Road, go .9 mile south on Ali' i Drive. Turn right on Kaleiopapa Street at the "dead end" sign. Drive on to the "Keauhou Pier" sign. Park outside the tiny congested pier parking lot if you don't want to get stuck in a mini traffic jam. It's a short walk in from the bigger lot.*

We review boat trips departing Keahou Harbor at the end of the beach section (*see page 82*).

HE'EIA BAY

He'eia Bay is also known as Walker Bay. You can snorkel here without worrying about boats; however, entry is somewhat tricky, since it's fairly shallow and more rocky than sandy. Since the bay is long and narrow, it can be quite calm if the wave angle is OK. Snorkel anywhere within the bay. No facilities are available. There's a beach, but it's not the sandiest or prettiest. We'd call this a good place for snorkeling, but not as good a prospect for swimming.

This tiny narrow bay located just north of Keauhou Bay (an easy walk) is quite protected and uncrowded. It's hidden in a small residential area just south of the Kanaloa development. There is public access from Manukai Street, although it's a bit hidden and overgrown.

GETTING THERE   *If you happen to be staying near Kanaloa, He' eia or Walker Bay is just a short walk south. If you're driving, take King Kamehameha III Road from Highway 190 down the hill past Ali' i Drive, and turn left on Manukai Road. Watch on the left for a small street.*

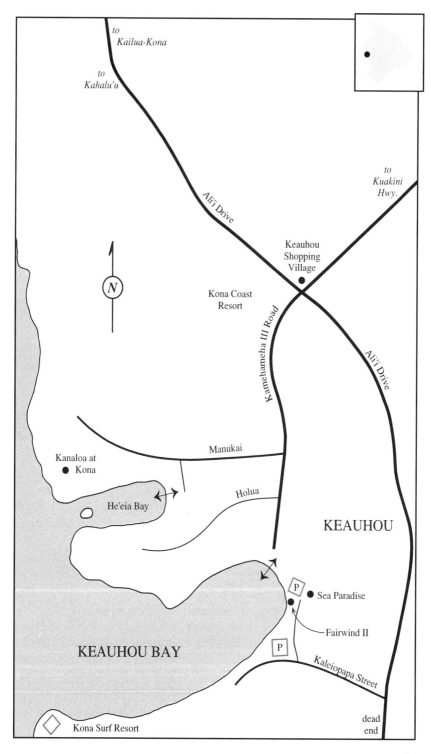

to
Kailua-Kona

to
Kahalu'u

to
Kuakini
Hwy.

Ali'i Drive

Keauhou
Shopping
Village

Kona Coast
Resort

N

Kamehameha III Road

Ali'i Drive

Manukai

Kanaloa at
Kona

He'eia Bay

Holua

KEAUHOU

P

Sea Paradise

Fairwind II

KEAUHOU BAY

P

Kaleiopapa Street

dead
end

Kona Surf Resort

KEALAKEKUA BAY

Kealakekua Bay is the site of the Captain Cook Monument on the north, as well as the town of Napo'opo'o to the south.

The bay tends to be very calm even in winter, so with fins it's possible to swim the one mile across the bay (actually less if you walk to the far end of the beach). Snorkeling is all along the far side with the best variety right near the monument.

The middle of the bay is just sand, not coral, but usually has about 50 spinner dolphins playing around in groups of 2 to 5. It's quite a thrill to watch them swim under you and jump out of the water. Snorkel the superb clear waters near the Captain Cook Monument, where you'll find coral, eels, turkeyfish, and an endless variety of colorful fish.

You can snorkel along the rocks back to the right until you exit at the beach. This is a long swim, so you need to either be in good shape or allow plenty of time to rest along the way. Fins are essential and a wet suit makes it easier to stay as long as you like.

Late afternoon might not be the best time to try this swim. The sea is often rougher then, and sharks are reputed to come in to feed as night approaches. Not everyone wants to be out in the middle of a large bay and see a shark up close.

It's also possible to see this area by kayak, which can be rented near the parking lot. If you do rent a kayak, make sure to enter the water from the pier area, rather than trying rough spots where you risk turning over on the rocks. This would be a perfect opportunity to bring a small picnic in a water-tight container, because you can go ashore at the small pier in front of the monument.

GETTING THERE   *Drive south on Highway 11 to about the 111 mile marker, watch for a Y with a small sign and a Hawai'ian warrior historical marker at Napo'opo'o Road. Head right, down the hill.*

Some books suggest hiking to the monument from here, since few vehicles can handle the rugged, steep road. However, it's about a 1,500' drop to the water, the road is rough, indirect and unmarked, and there's not enough shade and no water available, so think long and hard before trying this. We talked to one couple who desperately tried to buy a ride up from the bottom with no success. It made for a bad first date.

It's a long, windy drive down the road to Napo'opo'o, but it's well worth it, as you wind through some strikingly beautiful greenery and flowers.

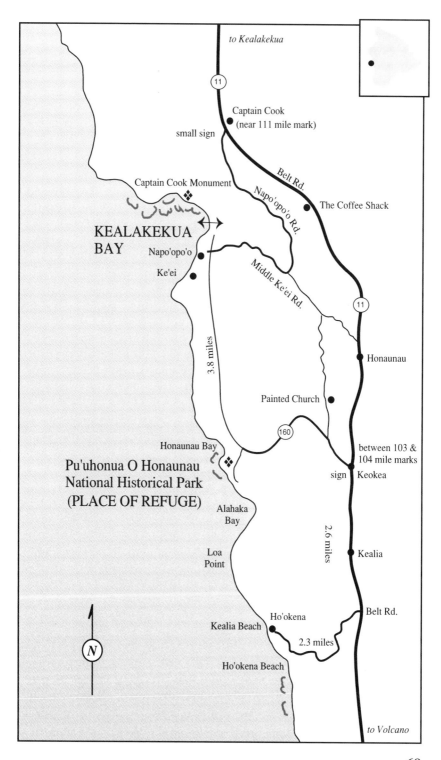

to Kealakekua

11

Captain Cook
(near 111 mile mark)

small sign

Belt Rd.

Napo'opo'o Rd.

The Coffee Shack

Captain Cook Monument

KEALAKEKUA
BAY

Napo'opo'o

Ke'ei

Middle Ke'ei Rd.

11

3.8 miles

Honaunau

Painted Church

160

Honaunau Bay

Pu'uhonua O Honaunau
National Historical Park
(PLACE OF REFUGE)

between 103 &
104 mile marks

sign   Keokea

Alahaka
Bay

2.6 miles

Loa
Point

Kealia

Belt Rd.

Ho'okena

Kealia Beach

2.3 miles

Ho'okena Beach

N

to Volcano

69

At the bottom of the hill you drive directly ahead into the parking lot. To be a few hundred yards closer, it's also OK to turn right and park where the road ends, offering a nice view of the bay.

To the right you can see a rocky beach where entry to the water is quite easy. Shoes are needed because the beach has more pebbles and rocks than sand. It's possible to enter right at the end of the road, or walk to the far end and enter closer to the snorkeling area. It just depends on whether you'd rather walk or swim.

Just two boats have permission to anchor within the bay. This is certainly a much easier way to get there, although expensive. We tend to prefer finding our own way, but in this particular case, we highly recommend the Fair Wind II (*see review, page 83*).

Since the snorkeling is so superb, it's great to save your time and energy for this incredible spot, where two hours pass in a flash. If you're serious about snorkeling or even willing to try it once, don't even think about missing this opportunity. Try to spot some of the beautiful Potter's angelfish in the shallow areas near shore.

Zodiac trips also stop at several spots around the bay; however, they tend to stay just a short time in each location. This spot really merits several hours. My preference would be an all-day trip.

PLACE OF REFUGE

Pu'uhonua O Honaunau National Historical Park is also referred to as The Place of Refuge (*see map, page 69, and History, page 110*).

This whole area offers excellent and extensive snorkeling, but I'll mention three that I would highly recommend. Most snorkelers seem to think they can't snorkel in the park, so they go right at the fire hydrant just before the parking lot, then park along the road and enter the water from there.

It's possible to snorkel as far as you like to the left along the edge of the park passing boulders through 20-30' deep clear water where we saw lots of large and small fish, more turtles, eels, interesting coral and great canyons. Besides the gorgeous setting, we had almost the whole place to ourselves.

Our preference is to enter through the park. The park charges $1 per person at the information desk for parking and use of all facilities,

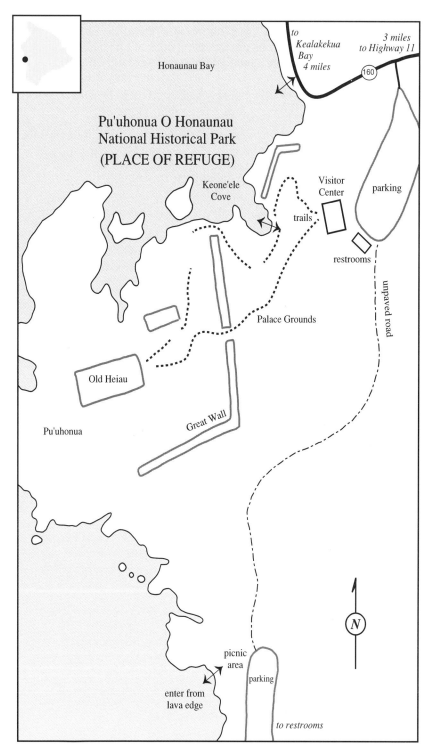

Honaunau Bay

to Kealakekua Bay 4 miles

3 miles to Highway 11

160

Pu'uhonua O Honaunau
National Historical Park
(PLACE OF REFUGE)

Keone'ele Cove

trails

Visitor Center

parking

restrooms

Palace Grounds

Old Heiau

Pu'uhonua

Great Wall

unpaved road

N

picnic area

enter from lava edge

parking

to restrooms

quite an excellent value. It's a fascinating and beautiful place that shouldn't be missed, so allow time to stroll around the grounds. The exhibits are interesting, and there's a well-marked trail. This was a religious site, and it does certainly have a serene and spiritual feeling. The information desk has an excellent brochure with map of the park and sells books as well as a large map of the Big Island.

Due to the historical character of the Park, beach users and snorkelers are not allowed to leave anything on the grounds (including towels). Just suit up at your car and walk through the beautiful, shaded grounds to the right to the little shallow beach (actually a canoe entrance). You may catch some odd looks from other tourists who wonder why you're here carrying mask and snorkel, but the folks at the front desk insist that it's perfectly OK. Swim out this old canoe channel and you're right in the middle of the best snorkeling. Several turtles frequent this area.

GETTING THERE   *From Highway 11 (see* map, page 71*), drive south about 7 miles past the Kealakekua turnoff. Take Highway 160 toward the water (between markers 103 and 104). It's an excellent road and well-marked. Near the water you'll see a sign on the left to the parking lot. Alternately, you can get here from Napo'opo'o by taking a straight one-lane road across the lava about 4 miles south. It joins Highway 160 right near the entrance to the park.*

Be a good guest

Remember that Big Island residents pay substantial taxes to provide lifeguards, showers (even when water is scarce) and clean up after wild tropical weather. You are given entirely free access to these fragile environments, while crowding the streets with rental cars, and certainly not contributing to the serenity.

Buying local products and crafts whenever you can is good for Hawai'i. Even more important, try to leave each beach a little better than when you arrived. If you can dive, use your skills to retrieve that soft drink can resting on the bottom. Anyone, even small children can pick up a bit of trash on the beach. This is something you owe to the people who live here and the people who will visit in the future. Help create the kind of environment you'd like to live in, and hand over to your children.

Another terrific spot is on the other side of the park in front of the picnic area. There's no beach here, just lava cliffs. It may not look easy to enter the water and isn't for beginners, but we found a fairly easy place. Park in the first section, and walk across the lava toward the water (shoes or booties are essential).

At first glance, entry would seem difficult, but look carefully and you'll find sloping shelves that make for practical entry and exit (use caution – this is for experienced snorkelers only if there is any swell (*see Understanding Waves, page 29*).

When there is little swell, getting out is easy, because you can look underwater for rock steps up. This is another wonderful and different site because the lava forms sharp canyons. In places the adventuresome can let the currents sweep them through holes and narrow canyons (only for the adventurous with snorkeling buddies, as well as good health and disability insurance). Snorkel in either direction. If you swim quite a ways to the right, you can watch fish considering whether or not to brave the fishing lines off the point. Water temperature changes are frequent here caused by the cool fresh water entering the sea.

GETTING THERE   *Drive to the left of the information building (*see map, page 71*). Instead of parking, continue through onto a small road past two signs that say "picnic" and "no parking". It angles to the right toward the water, but looks like it goes nowhere. Suddenly you'll come upon a parking lot and picnic tables, with restrooms and showers at the far end.*

This is another great picnic site. However, if you didn't bring a thing, you can always stop at the Coffee Shack up on Highway 11 on the ocean side about halfway between the Kealakekua turnoff and 160. Coffee Shack has a great view of the bay, gets cool breezes up the hill and has a nice variety of snacks from pastries to vegetarian pizza.

CHRISTMAS WRASSE

HO'OKENA

Ho'okena is one of several more distant beaches south of Kona and around the southern tip of the island. Most of these can be snorkeled only in calm weather, so ask about conditions locally before trying them. Snorkeling conditions are dependent on the direction of the waves and can sometimes be fine even during stormy winter months when wind direction changes – sometimes quite suddenly.

Ho'okena is a nice sheltered bay with snorkeling on the left. It does have restrooms, showers, palms, some tidepools and boat entry, although the facilities are fairly basic and not especially clean. Walking or driving to the right will take you to Kealia Beach which is poor for swimming, but good for snorkeling because of its wide, shallow coral shelf.

GETTING THERE   *Driving south on Highway 11 (see map, page 69), go 2.6 miles past the Highway 160 turnoff, which heads for Place of Refuge. The Ho'okena turnoff is well-marked, and so it's easy to find. Follow this paved road 2.3 miles toward the water at the end. It's only one and a half lanes wide and winds, but it's easy to drive.*

MILOLI'I BEACH PARK

There is a public beach on Miloli'i Bay. Just follow the signs to Miloli'i Beach Park. Snorkel here in calm weather only; there's lots to see right near the shore.

We've heard that Papa Bay just north has excellent snorkeling at the far right. We hesitated to try it because it's necessary to cross through the residential section to the right when you reach the bottom of the hill. Signs at each street to the right state "Private subdivision, not a public access, residents and authorized persons only, violators will be prosecuted". Local residents are serious about not allowing beach access, so watch out.

GETTING THERE   *On Highway 11 (see map, page 36), pass marker 89, then watch for the Miloli'i sign. The steep road down to the beach starts out as a newly-paved one-lane road, then becomes older and even narrower as it switches back and forth. It's about 5 miles down with some traffic (uphill has the right-of-way), so go nice and slow.*

WHITTINGTON BEACH PARK

Whittington Beach Park encompasses several lovely beaches with lots of picnic sites, facilities and tidepools. Most of the area is lava, so access is tricky, especially if there are any waves at all, which is most of the year. Plenty of wind hits this southern part of the island.

The best spot to enter the water is just left of the old wharf (at the far right of the park), where the sea tends to be more protected. Swim around the wharf to the right and snorkel the rocks to the right of this small bay. The entry here over lava and coral is definitely not for beginners – especially when there are any waves at all. Watch very carefully for the wind direction and currents. Ask local people for advice whenever possible; there are usually local families picnicking, or perhaps scuba diving groups.

Keep in mind that locals may find the place much easier than you do when there is swell because they've had more practice. It's not always as easy as you may think to keep from getting scraped (or worse) on the coral when a wave suddenly shows up (*see Understanding Waves, page 29*).

GETTING THERE   *As Highway 11 passes South Point* (see map, page 37) *it eventually comes close to the water at Whittington, which is along the right of the highway. As you come down a hill and see water near the highway, watch carefully. There is a sign, but it's set about 25' off the highway, so it's easy to sail on by. You have to make a sharp right at the sign and double back about 1/4 mile to the parking area.*

ORANGESPINE UNICORNFISH

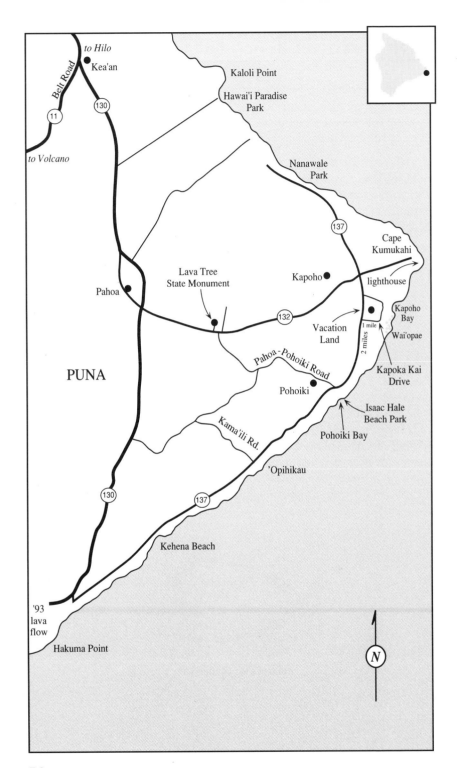

to Hilo
Kea'an
Belt Road
11
130
to Volcano
PUNA
Pahoa
130
'93 lava flow
Hakuma Point
Kehena Beach
137
Kama'ili Rd.
'Opihikau
Pohoiki Bay
Pohoiki
Pahoa-Pohoiki Road
Lava Tree State Monument
Kaloli Point
Hawai'i Paradise Park
Nanawale Park
137
Kapoho
132
Cape Kumukahi
lighthouse
Vacation Land
1 mile
2 miles
Kapoho Bay
Wai'opae
Kapoka Kai Drive
Isaac Hale Beach Park
N

ISAAC HALE PARK

On a calm day you can easily enter the water at the boat ramp (watching for boats, of course, which come through often in the summer) and snorkel straight out into a very protected area.

If the sea is close to flat, you can snorkel to the left around the pier to venture further out. It's not really difficult snorkeling, but winter requires fins if you really want to explore. This park is an attractive and popular place with lush surroundings to explore – including natural volcano-warmed fresh water pools in the lava. A small path leads off towards the right (facing seaward) to the pools.

GETTING THERE   *Take Highway 11 to Highway 130 (Kea' au-Pahoa Road). The intersection is at the town of Kea' au. Head south on Highway 130, a beautiful tree-lined drive with smooth pavement. At Pahoa, continue straight toward the water on Highway 132 through a lush overgrown area. Don't take the Lava Tree turnoff that you' ll see on the left.*

Just past the Lava Tree turnoff, hold to the right (holding left takes you to Kapoho) on Pahoa-Pohoiki Road, which eventually narrows to one lane. You' ll arrive at a wonderful large park, where you can park on the right near the boat entry area at the breakwater, where it's calmest. It's a very short walk out on the little breakwater if you want to check out conditions from the end. Snorkel from the sandy beach to the left around the breakwater as far as it is calm.

REEF
OCTOPUS

WAIOPAE TIDEPOOLS

The Waiopae tidepools area is at the Vacation Land development just south of Kapoho Bay – Kapoho means "depression" in Hawai'ian. This is a magical and unique place where you can snorkel in large interconnecting lava tidepools with unusual and very colorful coral with lots of smallish fish. The area is entirely protected by a long natural lava breakwater.

In the tidepools, you'll see coral quite unlike anything you see elsewhere on the Big Island, so go out of your way to get to this fairly remote spot. It's a long way from Kona, but worth the drive.

A local resident told us that visitors are quite welcome as long as they do not litter or damage their precious tidepools, which are still in excellent condition. We urge you to heed this local sentiment, and treat the tidepools with care. The tidepools are completely protected by a large reef, getting no swell or currents other than the tide coming and going. Any junk that goes in, tends to stay in. "Take only pictures, and leave only footprints" would be a good idea here. Taking a few pieces of litter with you on the way out would be even better.

There are no facilities at all, so come prepared. Walk out over the lava wearing shoes (no flip-flops) – this is fairly rough lava and you will need to walk about 2/3 of the way toward the breaking waves, about two hundred yards. The first tidepools you see are quite shallow, but as you proceed they become deeper and more interesting. Ask other snorkelers which pools are their favorites.

The best time to see the tidepools is at high tide, because it's easier to swim from pool to pool then. Check which way the tide is going and follow it – rather than swimming against it. The tide can help you glide from one pool to another over a shallow area.

Reef shoes or booties are better than fins here because they enable you to snorkel one pool, and then climb over to another – which is especially useful if the tide is low. (Any kind of old or plastic shoes are better than nothing). Gloves would be nice too, but we didn't see anyone who had any.

Most people were wearing fins, which are not needed since the pools aren't huge (although some are bigger than swimming pools). It's a little slippery getting in and you need to watch for a few sea urchins, but once in, the snorkeling is easy and captivating. The coral is green, pink, and purple – the most colorful and unusual we've seen in Hawai'i. The water is clear, calm and only about 5' deep so you get a wonderful chance to examine the coral up close.

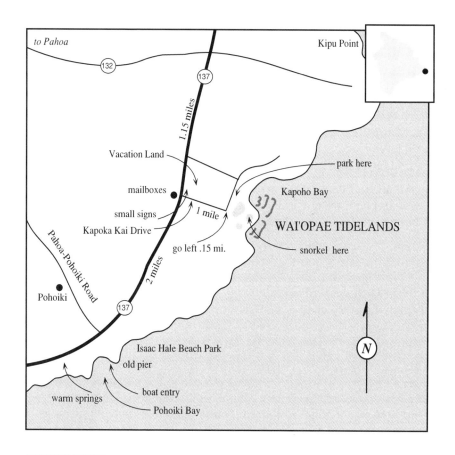

GETTING THERE   *From Isaac Hale (see* map, page 76*), take the beach road, Highway 137, north about two miles until you see a cluster of mailboxes on the left. A dirt road on the right (Kapoka Kai Road) is just before the mailboxes. Take a right here at a little sign that says "Vacation Land" and "private road". Continue on this gravel road one mile to the end. When the road curves 90° at the water, take a left on an even smaller road (Wai'opae Road) for .15 mile until you can see the tidepools. Most folks park here under the "no parking" signs along with a few other cars.*

From Kapoho: take 130 and follow the signs to Kapoho. Turn on Hwy. 132 just past the Lava Tree exit. Kapoho also has excellent tidepools, but is private and has actually posted a guard, so you need to head south along the beach road (137) 1.15 miles until you see the cluster of mail-boxes on the right, then take Kapoka Kai Rd. (a dirt road with a little sign that says "Vacation Land" and another "private road") to the left for 1 mile to the end, then a left on an even smaller road (Wai'opae Road) for .15 mile until you see the tidepools and park next to the no parking signs. The little town of Pahoa is a good spot to stop for snacks.

RICHARDSON BEACH PARK

Many guides say that there is no snorkeling near Hilo. Not true. There is a row of similar shallow beaches very close to Hilo – all of them along Kalanianaole Street (Highway 137) and very close together. Of the bunch, Richardson Park is definitely the best choice.

Snorkel near shore where entry is fairly easy over a tiny "beach" (*see Getting There below*). You can snorkel either direction. The near-shore area is well-protected by the outer reef, where the surfers play. In Hilo you may get to snorkel in the rain; think of that as picturesque, and rain snorkeling can be lots of fun.

The water can be a bit cooler here due to fresh water springs common at some Big Island snorkeling sites. Patches of fresh water are also the cause of the patches of "oily"-looking water (*see Doctor my eyes, page 51*). Just keep swimming and the temperature and clarity will often change. Visibility tends to not be as good on this side, due to the heavier swell action and rainwater runoff. If you hit a patch of weather that turns this side flat, and you are over here anyway, give it a try.

This is not a beach for people who like to keep some space between their body and the coral. Most of it is quite shallow – especially at low tide. There's actually enough clearance, but not for the claustrophobic. The day we snorkeled here it was so uncrowded we had the lifeguard to ourselves for awhile. He enjoyed telling us more about the many fish we had seen and seemed genuinely happy to have customers.

If you just want a cooling dip, the other nearby beaches are a better bet. The tidepools look inviting, but some are no longer clean, and have posted signs warning you to keep out.

A refreshing swim and shower are welcome – especially if you've driven all the way from Kailua-Kona, so don't let a little rain keep you from trying this spot.

GETTING THERE   *Cross the island to Hilo (see map, page 5). From the corner of Highway 11 and Kalanianaole Street. (the northwest corner of the Hilo Airport), take Kalanianaole Street. east along the shore. Drive exactly 3.7 miles, passing Reed Bay, Puhi, Onekahakaha, Leleiwi, all with lots of tidepools and shallow reef.*

Park in the little Richardson parking lot and walk toward the water. Showers and restrooms are available on the left close to the parking lot. Just behind the house continue walking to the right until you see the hidden lifeguard station. It's not far at all, just hidden by the vegetation.

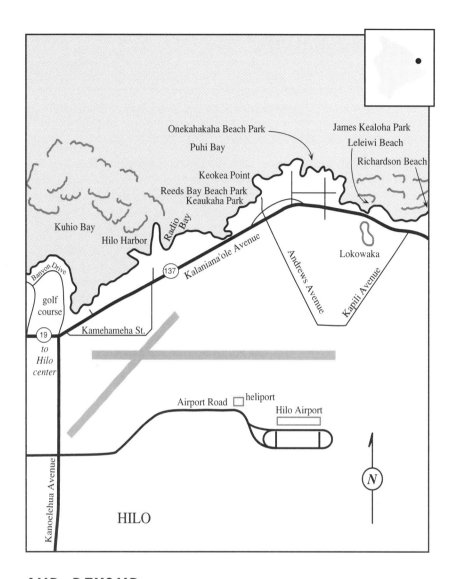

AND BEYOND...

This list of snorkeling beaches of the Big Island is certainly not complete, but we've tried to focus on the most accessible and skipped some of the ones where wave conditions are not as dependable – especially in the winter or during storms. We've heard that even the Waipio Valley beach has incredible snorkeling, but it's rare to have just the right conditions. The valley is also hard to reach by boat or land. Waipio is a spectacular valley, and certainly worth the long trip, but it shouldn't be considered a snorkeling destination most of the year.

WATER EXCURSIONS

There are enough good shore entry snorkeling sites on the Big Island to keep anyone busy for weeks. You certainly don't have to go on excursions to have a great time in or around the water, but you might want to, just to crank up the volume.

There are some great excursions available that give you access to places and experiences available no other way, or easier access to some great spots. Some folks find them the highlights of their Hawai'ian experience. The better excursions can add a lot of party spirit to your day, in exchange for a modest fee. If you ever get seasick, see *Motion Sickness, page 19*, for solutions.

It's hard to walk a full block in town without coming across a stand of free tourist publications. These colorful booklets are eager to help you find your way to particular opportunities, often with special discount coupons and offers. Pick up a couple at the airport on your way in, peruse them, and you'll soon be up to speed on the latest offerings.

If the prices seem too high for your budget, don't despair. There are often discounts available that can help bring some excursions within your reach. If you belong to any of the following categories, ask the excursion operator if you are offered special rates:

- Children. Often 6-17 years old. Sometimes under 12 years.
- Young Children (5 and under) are sometimes free.
- Seniors
- Repeat visitors. If you repeat excursions, some give an extra 10% discount.
- Kama'aina (Hawai'i state residents with local I.D). often get local rates.

Here are a few offerings that we've personally checked out, and find especially worthy of note. Have fun! Check out our web site, http://www.wp.com/snorkel_hawaii for updates (*see Snorkeling the Internet, page 122*).

PETROGLYPH OF CANOE

FAIR WIND II

The premier way to get to Kealakekula Bay, perhaps the finest snorkeling site in all Hawai'i (*see map, page 67 and review, page 68*).

Fair Wind II is a thoughtfully designed replacement for the original Fair Wind, destroyed by hurricane Iniki. A deluxe 60'x30' catamaran with a large covered deck offering lots of shade and cushioned seats inside, as well as ample sunning room on top. A nice combination of comfort and style. Rated capacity is 148, but policy is to take no more than one hundred passengers per trip.

Fair Wind II is based at Keauhou Pier, (*see map, page 67*), steps away from its office. Trips go out just about every morning, and some afternoons, too, subject to weather and demand. Captain Glen Craig says they miss only a few days a year due to weather. February and March tend to be a bit rougher trip, but Kealakekula Bay itself is almost always calm. Only two boats are allowed to moor here.

The boat leaves Keauhou at 9:00 a.m. every day of the week, and serves Kona coffee, local fruit and pastries on departure. After a pleasant and pretty 45 minute cruise south along the coast, you pull in to Kealakekula Bay, and moor near the Captain Cook Monument.

The staff distributes equipment on the way, making sure everyone has the appropriate sizes and a good fit. The equipment is good quality and varied, including prescription masks, inner-tubes and box viewers for non-swimmers, and a water slide for kids. Everyone is cautioned about not touching any of the coral, and then it's into the water. You find a show there that will dazzle anyone.

After an hour or so of snorkeling wherever you please, a hefty barbequed lunch is served, with regular or veggie hamburgers, salad, fruit, chips and more, fortifying you for another session of snorkeling. And you WILL want another session! There is time for another 45 minute snorkel before departure.

If you'd like a video of the trip, Jolene Versailles is all over the place above and below water, hard at work catching everyone in the show. She creates a good quality, fun record of the trip, set to catchy local music. She takes orders after the trip, $25 per video, and you can pick your video up at the Fair Wind office the next morning or have it mailed home.

Another interesting option for $40 extra on the Fair Wind is Snuba, with Lynn Ekstrom. If you're curious about scuba, but hesitate to invest the effort and money to get yourself certified, this is a great opportunity to see how much you like the underwater world (*see Snuba, page 84*). Lynn gives a fifteen minute training course on board, then takes 4-6 people at a time on a 45 minute underwater tour while others are snorkeling overhead.

This is a family business that has been around for more than 20 years. Rather than just talking family values, they live them. The employees come across as happy, well-trained, and hard-working. The boat and gear are all first-rate. This is one of the top excursions in the Hawai'ian Islands, in our opinion.

Adults $68; children 6-17, $38; under 5, free. The price is a bit higher than some other excursions, but this experience seems worth it. The afternoon trip, from 2:30 to 5:30 p.m., costs a bit less, as it is shorter and serves less food; they don't always go out, though. Everyone we talked to was glad to have gone on the longer morning trip.

The age range on our trip was from 2 to 80 years old. Children are encouraged to come along, as well as handicapped (access is OK to all facilities). This is one of our favorite excursions in Hawai'i – whatever else you do, don't miss going to Kealakekua Bay! 322-2788, 800-677-9461

SEA PARADISE DIVERS

Located at Keauhou Bay, next door to Fair Winds II (*see map, page 67*). Sea Paradise Divers is primarily a dive operation, but they're quite willing to take snorkelers along.

One of their specialties is night diving with manta rays off the Kona Surf Resort, just five minutes away. The rays can be seen from the hotel, but access to the water is far too dangerous right there, with large waves breaking against the lava cliff. Given the darkness and big swells that roll through here, we'd recommend coming by boat.

Although the Sea Paradise people insisted that diving was best by far, Mel dove, while I snorkeled, so that we could compare. We agreed that there was surprisingly little difference in this case. In fact, I appreciated having the surface to myself, while diving gridlock set in down below. Both of us were lucky and had thrilling, up-close encounters with huge (8' wingspan) mantas that were feeding on plankton by doing somersaults directly in front of our flashlights.

We were told to not touch the rays, which is good advice, better for their health and yours. The largest ray swam closer and closer to me, apparently having missed the briefing. His open mouth was bigger than my head and I could look into his mouth and clearly see his gills. Even though I knew he preferred to eat plankton, it was still an eerie

Snuba

Snuba was developed as a simpler alternative to Scuba for shallow dives in resort conditions. Because snuba divers are strictly limited in depth and conditions, and always accompanied by an instructor, training takes just 15-30 minutes. Two people share a small inflatable raft, which holds a Scuba air tank. A 20' hose leads from the tank to a light harness on each diver. A soft, light weight belt completes your outfit. Very light and tropical!

Once in the water, your instructor teaches you to breathe through your regulator (which has a mouthpiece just like your snorkel) on the surface until you're completely comfortable. You're then free to swim around as you like (only down to 20' deep, of course). The raft will automatically follow you as you tour the bay.

It's that easy! You have to be at least eight years old, and have normal good health. Kids do well, and it is so simple that senior citizens often give it a try.

We are certified scuba divers, yet we tried Snuba because this was a perfect place to see what made it special. It actually has some advantages over scuba in that you're free of the cumbersome equipment. There's none of the macho attitude you see sometimes on dive boats. Snuba strikes us as a fun, reasonably safe experience if you pay attention and use it according to directions. Where the reef is shallow, and conditions calm, it can be better than diving because you're so unencumbered in the water.

Warning: *pay attention* to the instructions, because even at these shallow depths, you must know the proper way to surface. You must remember to **never** hold your breath as you ascend, or you could force a bubble of air into your blood. Breathing out continually while surfacing is not intuitive, but absolutely necessary when you're breathing compressed air. This is especially important to remember if you're used to surface diving where you always hold your breath. Dive safely!

feeling, especially when he rubbed me with his body. In general, I prefer observing the food chain, rather than being part of it.

Another snorkeler had his mask knocked off by too close an encounter, and lost a contact lens. Not the best place to deal with this, but easier for a snorkeler on the surface than a diver 20' down!

Sea Paradise is a dive boat, rather than an excursion, with minimal staff, and it was a little too chaotic for our taste. If you're snorkeling in this kind of situation, keep in mind that you have to take care of yourself and stay clear of any dangers – mostly the human kind.

Dangers have this way of sneaking up on you. What gets you is what you least expect. The sight of a large manta on the surface prompted an over-eager passenger (with perhaps too much alcohol) to jump off the side of the boat. In spite of my super-bright dive light, she managed to jump on top of me. Ouch!

Although the weather was calm and we were only going a short distance, there was quite a swell coming in to the deep water off the point, and several divers became seasick. With the dark night, and the large swell, it was a bit disorienting on the dark boat. The boat lurching up and down made climbing back on tricky, too.

Despite some shortcomings, and minor injuries, we'd go again in a minute, because the mantas are so interesting. This is not a trip for beginners, though, as you may have figured out by now. Wetsuits are essential for night diving, and we'd recommend them for night snorkeling. All the dive shops have plenty available and Sea Paradise managed to fit everyone – more or less. A perfect fit isn't necessary in Hawai'i, since the water is fairly warm (*see Wetsuit, page 17*). Night dive, $70 per diver; $40 per snorkeler. 322-2500.

MANTA RAY

BODY GLOVE

This stylish boat anchors at the Kailua-Kona Pier (*see map, page 61*). Body Glove uses a van parked on the pier for its office, which is actually quite convenient. Free parking for the pier area is up a little alley from Ali'i Drive, just north of the old church. It's marked by a small green sign with a large P. Walk to their van on the pier at 8:15 a.m. for check-in, the boat leaves at 8:30 a.m..

Body Glove takes a fairly short trip north to a site just beyond the old airport, so travel time is short, unless they detour to see whales (or, in our case, a sinking ship). This is a well-organized boat with an charge) and snorkeling, and has plenty of good equipment. A continental breakfast is available on the way out and sandwiches, salad and drinks at the site. (Drinks other than coffee, juice and water cost extra on most of the boats). You can snorkel an hour or so, then eat enough to renew your energy for another snorkel after lunch.

Sites change with the weather, but one of their standbys is an area called Papawai (sometimes Pawai) that stretches north of the Old Airport runway. There's about a 30' drop-off, so snorkelers can view the ledge as divers glide by along the bottom. When calm it's fun to cruise the top of the coral and rocks in the shallow water near shore, but the surge made this tricky when we tried it on a high-swell winter day. It was fun, but the surges carried us back and forth rapidly, so most of the snorkelers chose to stay along the ledge where it was fairly calm. Since it's a short trip, we had plenty of snorkeling time and thoroughly enjoyed the sights as well as the friendly crew.

If the weather is uncertain, call first to make sure the trip is still going. Body Glove rarely misses a trip, however. If they do cancel, check the excursions heading for Kealakekua Bay, where swells are rarely a problem. Adults, $54; age 5-17, $24; under 5 free. 326-7122

Be Careful Out There!

PADI and NAUI attempt to regulate the diving industry with strict rules, since there are serious risks involved. No one is allowed to dive without certification (a "C" card) or at least phone verification of status. Anyone who wants to dive without proper training is certainly a fool, and the shops who will take such rash people out are equally foolish. But it happens every day.

We have seen excursions all over the world offering to take people down without proof of certification. This is not the mark of the highest level of safety consciousness, so keep in mind that other advice and services from such operators may be similarly casual. Always take extra care with any rental equipment.

Especially when business is slow, it's hard to resist taking divers (or snorkelers) to sites they can't handle. At least on the good snorkeling excursions, they usually keep a close eye on all their charges, so it can be like having a lifeguard along.

With a dive boat you may find yourself on the surface as a snorkeler in much rougher conditions than the divers 60' beneath you. Be your own lifeguard!

KONA COAST DIVERS

Jim Robinson provides a well-organized and well-stocked store with good ideas for fixing equipment or solving problems. Their van leaves for their dive boat at about 8 a.m. and returns about 1:15. Some dives are more appropriate for snorkelers than others, so check first about the location.

Dive boats are much smaller than the boats catering to snorkelers and usually require greater skills on your part because they may anchor in less protected sites. For experienced snorkelers, Kona Coast Divers offers a chance to try different locations when the weather is nice. It's also a smaller, more personal experience. In bad weather, the dive and snorkel boats are often all lined up along the few calm sites, so the main difference between excursions is then size and comfort of the boat, food and cost. $70 per diver; $30 per snorkeler, including lunch.
75-5614 Palani Rd, Kailua-Kona
329-8802, 800-KOA-DIVE

ATLANTIS

This is a submarine excursion, rather than a snorkeling experience. It gives an opportunity to check out the underwater world without getting wet. Lots of people are attracted to the exotic notion of going down in a real submarine for the first time. Ooga! Ooga! Dive! Dive!

The trip starts at the Atlantis office in the shopping wing of the King Kamehameha Hotel (*see map, page 61*). Free validated parking is offered in the hotel lot for the 2 1/2 hours needed.

Your group assembles in the office, and then takes a boat out to the submarine. We went down to 108' although a lot of the fish were seen at a depth of about 30-60', where 2 divers fed fish on either side of the sub to make sure everyone had a good view.

The commentary is educational and enjoyable. Being down 100' just isn't as different as you might think – other than the color loss at greater depths; but you do get a taste of the deeper diver's world. The ride is smooth and quiet.

Though we prefer getting wet, this is lots of fun. Our only complaint is that the windows are very low and the seating somewhat cramped and hard, which can be hard on your back. For kids, the windows are just the right height. If you're inclined to seasickness, take whatever you use to avoid it (*see Motion sickness, page 19*). Look up as you ascend at the beautiful changing texture of the surface of the water. $79 each for adults. Selected times, $69. 329-6626

NAUTILUS II

Nautilus II has an office right across the street from the Kailua Pier. It offers a less expensive fish-viewing alternative. A semi-submersible, the Nautilus II is essentially an elaborate glass bottom boat. It simply has lots of big windows that extend below the surface. Much fish life is shallow, and you do get to see many of the same creatures. The Nautilus has some advantages over the Atlantis, with its much larger windows allowing comfortable viewing sideways toward the reef and down toward the bottom. Views of fish start immediately with fish and some coral right there at the pier.

On the morning we went out, Nautilus II was planning to make the trip for just two passengers – which is impressive. It seems that the majority of the boats (and helicopters also) will cancel at any point if they don't think their passenger list is large enough to be profitable. So if some people don't show up, you may get dumped at the last minute; not fun when you've gotten up early to make a 7 a.m. departure. When we tried Nautilus, the crew cheerfully set out with 4 people on board without even cancelling the Snuba diver who went along to feed fish and bring numerous creatures to the windows so that we could have a close look. Most companies cancel in this situation, sometimes telling you that there are weather or mechanical problems.
Adults, $40; under 12, $35; Seniors $34. 326-2003

SAILFIN
TANG

If you love the reef...

- Show respect for the reef creatures by causing them no harm.

- Do not touch the coral, for that damages it.

- Come as a respectful visitor, rather than as a predator.

- Leave the many beautiful creatures you find there in peace, so that others may enjoy them as you have.

- Think of the creatures of the reef as fellow travelers in our life journey, and then you may comprehend their magnificence.

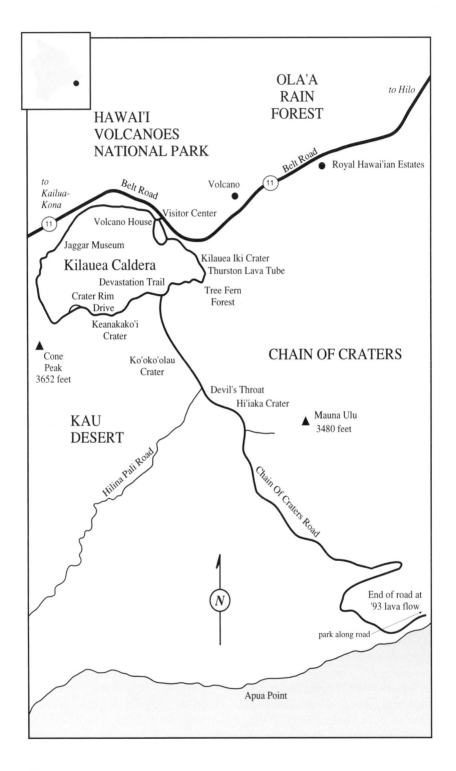

HAWAI'I
VOLCANOES
NATIONAL PARK

OLA'A
RAIN
FOREST

to Hilo

Belt Road

Royal Hawai'ian Estates

*to
Kailua-
Kona*

Belt Road

Volcano

11

11

Visitor Center

Volcano House

Jaggar Museum

Kilauea Caldera

Kilauea Iki Crater
Thurston Lava Tube

Devastation Trail

Crater Rim
Drive

Tree Fern
Forest

Keanakako'i
Crater

▲
Cone
Peak
3652 feet

Ko'oko'olau
Crater

CHAIN OF CRATERS

KAU
DESERT

Devil's Throat
Hi'iaka Crater

▲ Mauna Ulu
3480 feet

Hilina Pali Road

Chain Of Craters Road

N

End of road at
'93 lava flow

park along road

Apua Point

HAWAI'I VOLCANOES NATIONAL PARK

Once you've made it to the Big Island, you're very close to one of the most active volcanoes in the world. It's easy to get to the park, making for a great day trip. Massive active flows of lava all the way into the ocean are not uncommon. In January of 1983, Kilauea came to life, and has been active with scarcely a pause ever since. No one knows how long the island-building lava flow will continue. It certainly makes for a great show.

The upper slopes are a beautiful and fascinating location (although often quite cold when the clouds blow in) with much to see from fern jungle, to lava tubes, to moonscape with steaming vents. The seaside lies in a very different climate zone. There are 13 recognized climate zones, and Hawai'i, surprisingly, has 11 of them!

Stop first at the visitor center to watch the excellent video, and hear the day's weather and volcanic activity report from the ranger. Get a map, and catch a snack at the restaurant (Volcano House) overlooking the caldera, if you're in the mood.

Even the gift shop, the Volcano Art Center, is excellent. With a large selection of local crafts and Hawai'i books, it also has better prices than many shops around the island. There is usually a crackling fire going to warm chilly tourists who have come up from the beaches in shorts.

Crater Rim Road is a good starting point for a driving tour. Although the name sounds like it will follow the rim, this road also proceeds right through part of the caldera, with many interesting places to stop.

For hiking trails, consider Kilauea Iki (4 miles long) or Thurston Lava Tube (a very easy hike and nearly level). All trails and lookouts are well-marked and maintained. Maps are available at the visitor center.

Volcanoes Park helps protect many species of plants and animals endemic to Hawai'i, and you can do your part by taking nothing but pictures, and leaving nothing but footprints (and none of those in hot lava, please!)

After visiting the upper park area, take the time in the late afternoon to head down Chain of Craters Road toward the sea. It's a smooth, well-paved road (about 3/4 hour to go 25 curving miles) with parking at the end along the roadside. Walk along the day's marked path as close to the lava flow as you are currently allowed. There are dramatic new black sand beaches below the cliffs.

If you're lucky, you may be allowed to walk surprisingly close to the action – close enough to melt the soles of your shoes. We happened to arrive one day when lava was flowing quite near the road, and we were allowed to walk as close as 3' from the red-hot lava. Most impressive! Take a snack and drink, but only items that can take lots of radiant heat. If you stay until dark, you may see the hillside gradually turn bright red-orange with flowing lava.

Rangers are on hand to answer questions, direct you toward the day's action and attempt to keep most of you from killing yourself. Considering the dangers, it's amazing that people are allowed to get as close as we did. We were allowed to be arm-distance from red-hot lava – with no fence at all. In fact, we could have reached out and burned our fingers if we chose to. A great country, right?

Wear layers, because you'll be peeling them off fast if the lava is flowing anywhere near you! Signs suggest the sulfuric steam is not the best for those suffering from various ailments – especially lung or heart illnesses. If in doubt, limit your exposure with a brief visit.

Even though it's a long drive back to most hotels, Volcanoes Park is worth the effort. If your schedule allows, you may want to linger for a few days in one of the lodges or bed and breakfast inns. If you day-trip, as most folks do, drive carefully, take your time and savor this incredible volcanic activity. Reservations or current volcanic information is available at 967-7311.

As you head for Kailua-Kona after a long thrilling day, you'll no doubt appreciate the effort made to keep Highway 19 in excellent condition with more Botts Dots than absolutely necessary. Maybe they got a quantity discount? It's like driving on a night-time airport runway! This good road is a real blessing, though, when you're tired. It's a two-lane, very curvy road, so don't treat it as a road rally and try passing everyone on the way. Relax and enjoy the trip.

LAND EXCURSIONS

MAUNA KEA

Skiing is occasionally possible when there is snow, but not at all a sure thing. Even when it does snow, keep in mind there are no facilities – no lifts at all.

For sight-seeing, the Mauna Kea road offers a chance to drive from sea level to the top of the world (almost) in just a few hours. Take warm clothes if you go, as it is usually windy and chilly on top.

Driving to the top also a perfect chance to experience altitude sickness, which is more likely when you ascend quickly to high altitude and overexert. If you develop nausea, headache or a cough, descend immediately until symptoms disappear.

WAIPIO VALLEY

This is gorgeous rain forest country, and a memorable historic remnant of the Hawai'i of the ancient Polynesian settlers. A spectacular lookout is located at the end of highway 240. It's also possible to take a tour into the valley from Waimea in special vehicles. The road is much too difficult for tourists to negotiate in rental cars.

GETTING THERE   *See Big Island map, page 4. From Kailua-Kona, take Palani Road as it angles north (becoming Hawai'i Belt Road). to Waimea (also called Kamuela). From the north, take Highway 19 all the way to Waimea. From Waimea, continue east on Highway 19 to Honoka'a and then turn left on Highway 240, heading back to the west along the coast.*

SOUTH POINT

This is the southern-most point in the USA. Take South Point Road from Highway 11. It's a very narrow road, rough, quiet and wind-swept. Most rental companies don't like to have you drive their delicate cars on this road. Some tourists ignore what their rental car companies say, and drive it anyway, without damaging their car. If you're one of these scofflaws, have fun, and don't even think about needing a tow or gas!

Once you get there, admire the austere panorama. There's not a whole lot else to do in this rocky, dry spot.

AIR EXCURSIONS

Hawai'i Helicopters

Their French-designed Astar helicopter is incredibly smooth. The view is fabulous from the two front seats; four people are seated across the back and the two in the center have less spectacular views. Seats are assigned by "weight distribution" on boarding, so there's no way to be sure where you will sit. Be prepared to tell any helicopter the weight of each person in your party.

Weather conditions and vog (*see page 103*) are terribly unpredictable, making last-minute cancellations frequent in this business. Seeing a gorgeous, clear day in Kona tells you little about weather conditions on the other side of the island. Always call to confirm.

Hawai'i Helicopters seems to make a very conscientious attempt to call you, but can't always reach people. It helps to arrive early and the wait is quite pleasant because you can browse through their wonderful selection of Hawai'i photo books while sipping some wine. If you ever get seasick, plan ahead (*see Motion sickness, page 19 for tips*).

Hanger 106 just south of the main airport buildings
at Keahole Airport.
329-4700, (800) 994-9099

'Io Aviation

'Io Aviation is located right next door to Hawai'i Helicopters at the south end of Keahole Airport. ('Io means Hawai'ian hawk.) 'Io has a Hawai'ian pilot who does great talkstory. Very friendly people running a nice small helicopter operation. 329-3031

Papillon Helicopters

Papillon Helicopters depart from their own helipad near the Waikoloa area hotels (you can see them next to the highway on the mountain side).

A northerly departure point is more convenient for those staying further north, but a fair drive from Kona in traffic. (It can be 45 minutes or more from the condos south of Kona). We're hesitant to recommend them because they seem to have a cavalier attitude about wasting the customer's time. Twice we were told to drive up for a trip

only to find they had miscommunicated and weren't going to fly. We can understand weather changing, but we're not as sympathetic about poor communication. Since the best visibility is early, people prefer the 7 a.m. flights. Papillon's phone rings at the Keahole Airport, where the person answering may be unaware of decisions made at the helipad further north. Insist on speaking directly with the helipad people.
Keahole Airport: 329-0551
Helipad: 885-5995 (near the Waikoloa resorts)

Island Helicopters

Island Helicopters flies from Hilo, which is actually much closer to the great views than the Kona coast. Starting closer saves flying time required to cross this large island, so consider them if you plan to visit Hilo. Some people even fly over to Hilo from Kona and insist they save money. Since the airport is so close to town, it's easy to see Hilo in the same day if you like. Several other companies also fly helicopters from the Hilo Airport. (800) 829-5999

Blue Hawai'ian Helicopters

(800) 786-BLUE, 961-5600

Safari Helicopters

(800) 326-3356, 969-1259

Big Island Air

Fixed wing sightseeing excursions. (800) 303-8868, 329-4868 Keahole departures.

UFO Parasail

Air or sea – you be the judge. At the Kailua Pier. 325-5836

MILLETSEED BUTTERFLYFISH

MARINE LIFE

The coral reef supports tremendous diversity in a small space. On a good reef, you've never seen everything, because of the boggling variety of species and changes from day to day.

In Hawai'i the reef coral itself is less spectacular than in warmer waters of the world. This is counterbalanced by the colorful and abundant fish, which provide quite a show.

There are excellent color fish identification cards available in bookstores and dive shops. We particularly like the ones published by Natural World Press. There are also many good marine life books that give far more detailed descriptions of each creature than we attempt in these brief notes.

FISH NOTES

OCTOPUS

Some octopuses hide during the day, but others will hunt for food then. They eat shrimp, fish, crabs, and mollusks – you should eat so well! Octopuses have strong "beaks" and can bite humans, so it's safer to not handle them. Some octopuses even have a poisonous bite. Being mollusks without shells, they must rely on speed and camouflage to escape danger. Octopuses are capable of imitating a flashing sign, or changing their color and texture to match their surroundings in an instant. This makes them very hard to spot, even when they're hiding in plain sight. They can also squirt an ink to confuse predators.

SHRIMP
Come in all kinds, colors, and sizes – and like to hide in rocks and coral. They are difficult to spot during the daytime, but at night you will notice lots of tiny pairs of eyes reflected in the flashlight glare. Most are fairly small and well disguised. Some examples include: the harlequin shrimp (brightly colored) that eat sea stars, the banded coral shrimp (found all over the world), and numerous tiny shrimp that you're not likely to see without magnification.

SEA URCHINS   Concealed tube feet allow urchins to move around in their hunt for algae. The collector urchin has pebbles and bits of coral attached for camouflage. Collector urchins are quite common in Hawai'i, and have no spines. Beware of purple-black urchins with long spines. These are common in shallow water at certain beaches. It's not the long spines that get you, it's the ones beneath. The bright red pencil sea urchin is common and easy to spot. Although large, its spines aren't sharp enough to be a problem for people.

SEA STARS   Abundant, in many colors and styles. The crown of thorns sea star, which can be such a devastator of coral reefs, is found in Hawai'i, but not in large numbers like the South Pacific. Sea stars firmly grasp their prey with strong suction cups, and then eat at their leisure.

RAYS

Manta rays use two flaps to guide plankton into their huge mouths. They are quite common at Big Island beaches. Mantas often grow to be two meters from wing-tip to wing-tip, and can weigh 300 pounds. Even larger specimens are sometimes seen by divers. They don't sting, but are large enough to bump hard. Mantas feed at night by doing forward rolls in the water with mouths wide open. Lights will attract plankton which appeal to the manta rays. Dive boats in certain locations can easily attract them with their bright lights making the night trips quite exciting. Another beautiful ray, the spotted eagle ray, is less common, but can sometimes be seen cruising the bottom for food and can grow to be seven feet across. Spotted eagle rays have a dark back with lots of small white dots. Common sting rays prefer calm, shallow, warmer water.

EELS

Eels in Hawai'i include the commonly seen moray eel – the most common types include whitemouth and snowflake eels. They can easily grow to two meters. Varieties of moray found in Hawai'i include the zebra moray (black and white stripes), wavy-lined moray, mottled moray, and dragon moray (often reddish-brown with distinct white spots of differing sizes). Morays prefer to hide in holes during the day. When they stick out their heads and breathe, their teeth are most impressive.

TRUMPETFISH

These long, skinny fish can change color, often bright yellow or light blue – and will change right in front of your eyes. They sometimes hang upright to blend with their environment, lying in wait to suck in their prey. Trumpetfish are quite common in Hawai'i and often seen alone. Some grow to more than one meter long.

NEEDLEFISH These slim,usually silvery-blue fish like swimming very near the surface, usually in schools. Needlefish are as long and skinny as their name implies, and grow to as much as 1-2' long. Color and markings vary, but the long narrow shape is distinctive and hard to mistake. They're usually bluish on top, and translucent below for camouflage.

BUTTERFLYFISH

Butterflyfish are beautiful, colorful, abundant and varied in Hawai'i. They have incredible coloration, usually bright yellow, or orange and black, with a little blue. They hang out near coral, eating algae, sponges, and coral polyps.

Butterflyfish are often seen in pairs because many mate for life. They're delightful to watch, fast and fascinating.

The ones you are most likely to see in Hawai'i include: raccoon (reminding you of the face of the animal), ornate (with bright orange lines making it easy to spot), threadfin (another large, beautiful one), saddleback, lemon (very tiny), bluestripe (a beautiful one found only in Hawai'i), fourspot, milletseed, oval, teardrop, and forceps (with its distinctive long nose).

An interesting one is the pyramid, which can darken its white body, then light up a spot in the center, making it look just as though it swallowed a light bulb. The lined butterflyfish is the largest found in Hawai'i.

Many butterflyfish have black spots near the tail – perhaps to confuse a predator about which way they're headed.

Among the most dramatically colored fish on the reef, male parrotfish are blue, green, turquoise, yellow, lavender, and/or orange with assorted variations of these colors. Females tend to be reddish brown. No two are alike. These fish change colors at different times in their lives and can also change sex. They can be quite large (up to 1 1/2 meters). Parrotfish are very beautiful, with artistic, abstract markings. They constantly scrape algae from dead coral with their large, beak-like teeth, and create tons of white sand in the process. Most prefer to zoom away from snorkelers, but you'll see them passing by.

TRIGGERFISH

Fond of sea urchins as a main course. Triggerfish varieties include Picasso (wildly colorful – not too many at each beach, but worth watching for), reef (the Hawai'ian state fish), pinktail (easy to identify with its black body, white fins and pink tail), black (common, distinctive white lines between body and fins). The checkerboard triggerfish has a pink tail, yellow-edged fins, and blue stripes on its face. All triggerfish are very beautiful and fascinating to watch.

FILEFISH

The scrawled filefish has blue scribbles and brown dots over its olive green body. Quite large, up to 40", often in pairs, but occasionally in groups. A filefish will turn its body flat to your view, and raise its top spine in order to impress you. This lets you have a great close-up view. The brown filefish (endemic) is much smaller, with lines on its head and white spots on its brown body. The fantail filefish (also endemic and small) has a distinct orange tail and lots of black spots over a light body. Filefish will sometimes change color patterns rapidly for camouflage.

SURGEONFISH

Razor-sharp fin-like spines on each side of the tail are the hallmark of this common fish. Varieties includes the orangeband surgeonfish (with distinctive long, bright orange marks on the side), as well as the Achilles tang (also called naso tang), which has bright orange spots surrounding the spines near the orange tail. The yellow tang is completely yellow and smaller. The sailfin tang has dramatic vertical markings. It's less common, but easy to identify.

WRASSES

Wrasses are amazingly bright and multicolored fish. Some very small ones set themselves up for business and operate a cleaning station, where they clean much larger fish without having to worry about becoming dinner. They eat parasites, and provide an improbable reef service in the process. Perhaps their bright colors serve as "neon" signs to advertise their services. In Hawai'i the cleaner wrasse is bright yellow, purple and black. Other wrasses are large including the dazzling yellow-tail (up to 15"), which has a red body covered with glowing blue spots, a few stripes, and a bright yellow tail. Another large wrasse, the saddleback, is endemic to Hawai'i. It is bright blue, with green and orange markings. Wrasses are closely related to parrotfish. Like parrotfish, they can change colors and sex.

SCORPIONFISH

Beware of poisonous spines! Don't even think about touching a scorpionfish, and try to avoid accidentally stepping on one. This varied group of exotic fish includes the Hawai'ian turkeyfish, sometimes called a lionfish. This improbable-looking fish is very colorful, with feather-like multicolor spines.

Others are so well-camouflaged that they are hard to see. They just lurk on the bottom blending in well with the sand and coral. If you see one, count yourself lucky.

PUFFERFISH Pufferfish can blow up like balloon when threatened. Two kinds are common in sheltered areas: porcupine (displays spines when inflated), and spotted (brown or black with lots of white dots). They tend to prefer to escape under the coral, although some seem unafraid.

WHALES

Humpback whales migrate here to breed in winter (around mid-December). Humpbacks come quite close to the coast, where you can watch whole families. They are so large that you can often see them spouting and breaching quite easily from shore. If you bring binoculars, you can see them well. Their great size never fails to impress, as does their fluid, seemingly effortless graceful movement in the water. Other whales often seen around the Big Island include: sperm whale, pilot whale, exotic false whale, pygmy killer whale and the beaked whale.

DOLPHINS Spinner dolphins are frequently seen in large schools. They swim as small family groups within these schools, and often swim fast, leaping from the water, spinning in the air. They tend to hang out in certain locations, so you can search them out if you like. Bottlenose dolphins like to approach fast-moving boats, and it is a great thrill to watch them race along just next to the bow of your boat, jumping in and out of the water with grace and easy speed. Beaked dolphins are also commonly seen in Hawai'i.

SEA TURTLES

Common at many beaches, though they usually stay away from humans. Some do seem nearly tame – or at least unconcerned about snorkelers. Sea Turtles are often seen in pairs. Large specimens (such as some at Kahalu'u Beach and Place of Refuge) might be more than 100 years old, and tend to be docile and unafraid. They will let you swim as close as you like, but if you hover over them, they may be afraid to come up for air. Try to not disturb these graceful creatures, so that they remain unafraid to swim among snorkelers.

WEATHER

All islands have a windward side, which is wetter, and a leeward side which is drier. In Hawai'i, the northeast is windward and hence wet, and the southwest is leeward, or kona, and hence drier and sunnier.

Hawai'i gets most of its rain in the winter. The most severe storms ("kona"), however, come from the south and can even bring hurricanes in the summer. Temperatures tend to be very mild year-round, yet there is terrific variety any day of the year. There are days when you could tan in Kona in the morning, drive up to Mauna Kea to ski at lunch, and use your rain slicker in Hilo that afternoon.

Evaporating moisture from the ocean forms clouds. As the clouds rise over the mountains, they cool, and the condensing moisture becomes rain. The average rainfall on the Big Island is about 25" per year. On the windward side of a mountain, though, there can be over 250".

Having lost most of their moisture in passing over the mountains, the clouds have little left for the leeward side – so it is in the "rain shadow" of the mountains. The leeward weather is therefore often sunny. Waikiki, Poipu, Kaanapali, and Kona are all rain shadowed. On the Big Island if you get stuck in heavy rains somewhere, just head for Kona to find the sun. Kona averages only 15" of rain a year.

It's easy to see the effect of the mountains on rainfall by driving up the Kohala Coast. You transition from arid dryness to lush greenery within two miles. At the Volcano Visitor Center tourists huddle near the fire in the gift shop, while those at the end of Chain of Craters Road are peeling off jackets and sweating as they hike over warm rocks to await the sunset.

Changeable is the word for Big Island weather – not just between areas, but also rapidly changeable in any given place. The trade winds blow about 90% of the time in the summer and about 50% in the winter. They tend to be stronger in the afternoon.

October through April brings Kona storms (from the south), with waves reaching 10-15' high. You'll see the surfers heading for the beaches. A southern swell can come from the southern hemisphere and bring usually small to medium 1-4' waves. A north Pacific swell in winter or early spring can also bring high waves of 8-15'. Summer hurricanes, however, produce the most severe weather.

Tides are very slight here, with the average difference between high and low only 2'- 3' max. It's a good idea to know which way the tide is going because tidal flow does affect the currents. If the tide is going out, you might want to avoid snorkeling in places where currents tend to sweep out of a bay, often the center, or a gap in the reef (*see Understanding waves, page 29*).

Water temperature on the surface varies from a low of about 72° F (22° C) in March to a high of about 80° in September. Sheltered bays and tidepools can be a bit warmer, while deeper water can be surprisingly cool. If you happen to be slender, no longer young, or from a moderate climate, this can seem colder than you might like. The Big Island does have the warmest water in Hawai'i.

HURRICANES   Summer is hurricane season, but it is also when the weather is usually excellent. The storms don't last long, but can be terribly destructive. Hurricanes can bring amazingly heavy rain and winds to all the islands. Any of the islands could receive a direct hit. The Big Island has escaped major damage from hurricanes in recent years, as the eye of the hurricane veered off toward the west each time.

VOG   Sulphur dioxide mixed with fog and smog produces vog. Kilauea produces enough vog to cause problems on all the islands, but it is most bothersome in Kona and the south coast. Vog usually isn't a big problem to visitors, but it is more serious for residents. Vog location depends on wind direction and speed, so it sometimes hangs over Hilo instead of Kona.

SANDSTORM   A Hawai'ian sandstorm can carry volcanic sand, which is sharper, and hence worse than coral beach sand. It can suddenly blow across toward the big hotels on the coast north of Kailua-Kona and last for days. A sandstorm will definitely keep you off the beach there, but Kona is usually just fine. In Kona you won't even have a hint that wind is a problem while people are hiding out inside on the North Kohala coast.

TSUNAMIS   Those imposing sirens you see up high on poles near the shore are for tsunami alerts. Huge waves can be triggered by earthquakes either locally or far across the Pacific. They've hit the Big Island numerous times, more often in the north. Some very destructive tsunamis have hit Hilo and swept over the southeast, where the land is quite low. Depending on the exact direction, they can directly hit a valley and really wipe it out and rinse it clean. It is probably better to not be there when this happens, unless you're one great surfer dude.

Currently there's plenty of warning and authorities prefer to warn of every possible tsunamis just to be safe. It doesn't pay to ignore warnings just because the sea appears calm. If a major earthquake strikes while you're visiting, it's a good idea to head rapidly for high ground. Leave bays or valleys which can act to funnel the effects of a large wave.

GEOLOGY IN THE FAST LANE

The Big Island is a geological happening place – earthquakes, volcanoes, tsunamis. Something is happening every day, sometimes too minor to attract attention, and other times too spectacular to ignore. Lava flows recently have been creating new coastline to the southeast, with dramatic black sand beaches. Kilauea has been constantly active in recent years, although sometimes in a much more exciting fashion. This is one island that people fly to, instead of away from, when a big eruption takes place.

To understand what's happening now, cast your thoughts back about 30 million years. At that time lava was bubbling out in the middle of the Pacific about 20,000' below the ocean surface, due to a volcanic hot spot directly underneath. Molten rock pushing up through the ocean floor formed volcanoes under the sea. Lava built up, layer after layer until it finally reached the surface to form the first island.

As the volcanoes grew, the weight of these early islands gradually caused them to sink down again, forming atolls. The Pacific Plate drifted northwest, while the hot spot remained stationary. A long string of more than 30 islands were formed, stretching from Midway Island southeast 1600 miles all the way to the Big Island. Another island is already rising in the sea close to the southeast side of the Big Island. Loihi Sea Mount is now just 3000' under the surface, and will probably join the Big Island as it emerges. Lava flowing into the sea from Kilauea has been intermittently building the Big Island daily toward Loihi.

Most of the current above-water mass is now concentrated in eight islands. Kaua'i, about 5 million years old, is the oldest of these, while the Big Island is less than 1 million years old. As these islands drift around 4" northwest each year, the lava conduits to their volcanoes bend until new conduits are formed. Eventually, the next volcano in the chain takes over the job of releasing the unremitting pressure from pools of magma far below.

The Big Island is where the visible action is today. The island is made up of five large shield volcanoes, with another one buried underneath. Kilauea and Mauna Loa are both still active and building the island daily toward the southeast.

Hualalai (rising high to the east of Kailua-Kona) last erupted in 1801. It is currently swelling up, and is considered capable (and perhaps highly likely) of erupting violently in the future. Mauna Kea and Kohala have been inactive in recent history.

AND A LITTLE NATURAL HISTORY, TOO

When each underground mountain emerges from the sea, coral larvae begin to establish their new homes on the volcanic rocks around the base. Stony coral is one of the first ocean creatures to reach and become established on a new island.

These larvae travel island by island – originally coming in a very round-about fashion on the currents from the ancient reefs surrounding Indonesia. Once they became established, it was easier for new larvae to reach the next nearby island. The reef begins as a fringe around the island. Each polyp of coral secretes a skeleton of calcium carbonate. Gradually the colony grows large enough to provide a home for other plants and animals.

All of the major Hawai'ian islands now have fringing reefs around much of the shore. The Big Island, still in formation, is not yet fully surrounded by reef. As the islands grow, get heavy and gradually sink, the reef changes as well. The older islands of Kaua'i and O'ahu have very old coral reef deposits on land – remnants of a time when the sea level was higher.

Coral reefs are made up of coral animals and algae growing on top of the dead skeletons of former creatures. In search of sunlight, they continue to grow upward toward the light, as they need to stay within 150' of the surface of the sea.

The outside of a reef grows faster than the inner surface, so eventually a lagoon forms between the reef and the land. The reef is then called a barrier reef, limited examples of which can be found at places in Kaua'i and O'ahu.

An atoll occurs when this reef surrounds the island and the island itself sinks out of sight, leaving a ring of land formed on the remains of the formerly submerged barrier reef.

Since the currents in Hawai'i come mainly from Japan rather than the warmer south Pacific, they bring less variety of sea life. Larvae need to survive long enough to reach an island and establish themselves before sending out the next generation, so it's helpful to have stepping-stone islands in order to have the largest variety. Most will not survive long enough to cross large open ocean.

Tahiti, for example, has a much greater variety of coral because of the stepping-stone islands leading all the way from Southeast Asia. Hawai'i, in contrast, is one of the most isolated island groups in the world. It also has somewhat cooler water and less sunlight than Tahiti, making it less hospitable to some species. This isolation has kept all species of plant and animal life rather limited, and also encouraged the evolution of unique species found only in Hawai'i. These unique species are referred to as "endemic". They give Hawai'i a special character – both above and below the water. More than 30% of the fish seen here are found nowhere else in the world.

For millions of years the Hawai'ian Islands had no plants or animals in spite of the rich soil, due to their 2000 mile isolation from other large land masses. When plants and animals finally did arrive, they had little competition and a superb climate. The lack of competition meant plants did not require thorns or other protective features. Some plants and animals found such a perfect environment that they thrived. Before man arrived, Hawai'i had no fruits or vegetables. The Polynesians, and later Europeans, changed this environment enormously by their imports and cultivation.

Most of the "exotic" plants that you may think of as quintessentially Hawai'ian were brought by man (mango, papaya, pineapple, orchid, ginger, hibiscus). Koa and ohia (the Hawai'ian state tree), on the other hand, pre-date man's arrival. Ohia is often the first to grow in lava flows and has produced much of the Big Island rain forest.

Unfortunately, most of the rain forest has already been destroyed by animals brought by man (such as cattle and goats) or cleared to provide land for sugar production. Sugar and pineapple production appear to be on the way out, a casualty of world economics.

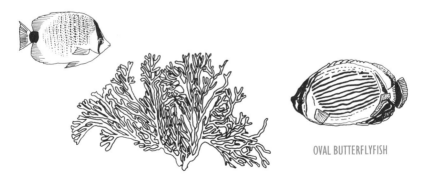

OVAL BUTTERFLYFISH

REEF DEVELOPMENT

Hawai'ian reefs have weathered at least four major changes in the distant past. Many land-based plants and animals also became extinct during these changes and others took their place. Current reefs are composed mostly of shallow water reef coral. They incorporate algae in their structure, and the algae is dependent on photosynthesis.

Different plants and animals live in different locations on the reef depending primarily on wave action. Species living on the outer edge of the reef are skilled at surviving strong waves and currents. Lagoon species don't have to endure this, and hence can be quite different.

Hawai'i has a number of strikingly different reef habitats – each with its own story to tell. Where the water is rough, cauliflower coral dominates. The more delicate finger coral grows only in the calm lagoon areas. Large boulders are common in the open waters, especially where wave action is heaviest, and they support entirely different creatures. Caves, caverns and lava tubes are quite abundant here. Steep drop-offs serve as an upwelling source of plankton-rich water, which attracts many larger creatures to feed. Sandy habitat is found more on the older islands to the northwest, with less on the Big Island.

Each plant and animal has a story to tell and there's much for the snorkeler to learn. There are many fascinating books that provide as much detail as you can handle. We'll just whet your appetite with a few brief notes.

Sponges, for example, are the simplest animals, really just a collection of cells. In Hawai'i they're found clumped on reefs.

Algae such as seaweeds produce oxygen and carbohydrates with the help of sunlight (a process called photosynthesis). In Hawai'i you'll see much red coraline algae.

Coral, sea anemone, jellyfish and hydroids all have simple body plans, but definite organs.

Flatworms are oval and free-living. Segmented worms are even more complicated (similar to earthworms).

Mollusks are protected by shells. They're called bivalves if they have two shells. Squid and octopus are mollusks, but have lost their shells.

Fish in the tropics can be extraordinarily colorful. Some can change colors rapidly, some blend into the background almost perfectly. Other fish are so oddly colored that they may confuse a predator. Colored spots near the tail that look like eyes is one defense mechanism, to confuse the predator. Often fish such as the cleaner wrasse use neon colors as a way of advertising their services. Other brightly-colored fish warn predators that they taste awful.

ABOVE THE WATER

Birds to watch for include nene, shearwater, albatross, myna, sparrows, pueo (Hawai'ian owl), stilt, coot, egrets, herons, doves, munia, cardinal, and finch.

Most of the "exotic" trees were brought in by man. Examples include mango, papaya, pineapple, orchid, ginger, and hibiscus. Koa, however, is native, and harvested to make excellent furniture. It was also used by the early Hawai'ians to make their beautiful canoes. The ohia, which is the Hawai'ian state tree, is another native. It often is the first plant to grow on lava flows. It's used for canoes, bowls, and other carvings.

Lowland rain forests in Hawai'i are mostly native ohia trees of all sizes. Most have been cut for cattle, agriculture, destroyed by animals introduced by man, or burned by flowing lava. There is still a bit of rainforest remaining on the Big Island, especially between Hilo and Volcanoes Park.

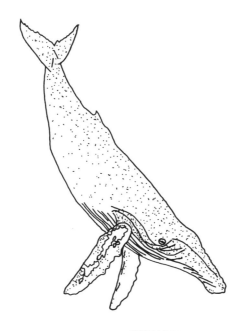

HUMPBACK WHALE

HISTORY

The islands of Hawai'i sat in dramatic isolation for millions of years, slowly softening their volcanic profile, while developing a soft green backdrop. The first people to arrive in the islands were from Polynesia, a culture with a long history of island hopping, dating back to their migration to Polynesia from the Middle East in large double hulled boats.

Fiji was settled by about 3500 B.C., then Samoa and Tonga, and later Tahiti. Hawai'i itself was first settled in waves beginning at least 1200 years ago, probably by voyagers from Tahiti and the Marquesas. They brought much of what they needed for a new life here: chickens, dogs, pigs, coconut, bananas. The first landing may have been on the Kona Coast. Many were apparently looking for a form of religious freedom and prospered in this new land. To migrate this far over open ocean required considerable planning and navigational skill, as well as strong motivation.

PETROGLYPHS

There seems to have been much interaction between the various islands of Polynesia and Hawai'i, followed by about 500 years of Hawai'ian isolation. Prior to Captain James Cook arriving in 1778 with two ships from England, there were perhaps 80,000 people on the Big Island.

Cook's timing couldn't have been better, or so it seemed at first. When he arrived, the Hawai'ians were celebrating a yearly feast in honor of the God Lono. When his ships arrived, he may have been thought to be the God in person, arriving for his feast. Captain Cook and his party were welcomed to the feasts and ceremonies. Local women swam out to greet the sailors. Cook reciprocated with a fireworks display. However, when one of his party died, it became obvious to the islanders that these were mere mortals after all.

When his two ships left Kealakekua Bay after two weeks of festivities, they encountered a storm that damaged one mast, so they returned. This time, their timing wasn't so good: the celebrations were over and the bay off-limits due to kapu (taboo). When they tried to land anyway, trouble started, leading to fighting, captive-taking, and eventually escalated until Captain Cook was killed.

There are several versions of this story, but all seem to agree that a major cultural misunderstanding led to the conflict, and Captain Cook's death. Driven into the shallow waters at the edge of the bay, he probably could have swam the short distance to the safety of a rescue boat. As with many sailors of that day, he had never learned how to swim, and was clubbed to death. A monument to him was built by the British. The only nearby buildable flat land remains under the control of Britain. The rugged isolation and lack of development of the north side of Kealakekua Bay has protected this magnificent snorkeling site to this day.

After Cook's death, King Kamehameha set out to bring all of the Hawai'ian islands under his rule. By 1791 he controlled the Big Island, by 1795, he had added Maui, Moloka'i, Lana'i and O'ahu. By 1810, he added Kaua'i, bringing the islands under one rule.

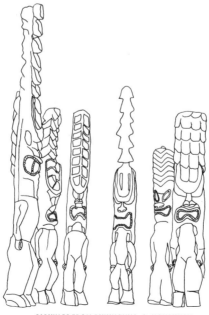

CARVINGS FROM PU'UHONUA O HONAUNAU

In 1820, Calvinist missionaries from Boston landed in Kona and immediately set about converting Hawai'ians, destroying their alters and building churches. They did put the Hawai'ian language on paper using the Latin alphabet as best they could. They also introduced the idea of private property. The missionaries, whalers and other newcomers introduced diseases like measles, for which the islanders had no antibodies, and these diseases decimated the population.

By 1860, immigration was actively encouraged by business interests in order to secure enough laborers for the sugar industry. These new arrivals came from all over the world: China, Japan, Philippines, Portugal, Russia, Spain, England, and America. Business ultimately came in conflict with royalty. Queen Liliuokalani was pressured by the American military to turn over control of the Republic of Hawai'i, which was annexed to the United States in 1898. Eventually the voters of the islands chose statehood in 1959.

The Big Island has become a center of education – particularly involving astronomy, geothermal and alternative energy, and ocean research. Agriculture plays a large role in the state's economy – with coffee, flowers, vegetables, fruit and macadamia nuts leading its many exports. Marijuana, though politically under fire and currently illegal, has been rated by some as the number one cash crop. Sugar is no longer profitable. Tourism is increasingly becoming a vital industry.

Topics of political interest on the Big Island recently include lively discussion of land use. A number of native Hawai'ians want to have land returned to their families, land which was taken by the missionaries, plantations and the military.

Sometimes land was simply made useless when water was diverted to the gigantic sugar cane fields, which were subsidized for so long until just recently closed.

Religious and business interests long had effective control of the islands, sometimes with the best intentions. The results, most people would agree, were sometimes good, and sometimes terrible. Hawai'i is engaged in a re-evaluation of its history, and its future – with a need for careful planning for the future of its people and their resources.

PETROGLYPHS OF BIRDMEN

LANGUAGE

English is now the official language of the islands of Hawai'i – except for the island of Ni'ihau. However, most place names and lots of slang are Hawai'ian, so it's most helpful to at least be able to pronounce Hawai'ian enough to be understood. It's a very straight-forward phonetic language – each letter is usually pronounced just one way. The long place names aren't nearly so daunting when you've learned the system.

All syllables begin with a consonant that is followed by at least one vowel. When the missionaries attempted to write this spoken language, they used only seven consonants (h,k,l,m,n,p,w) and five vowels (a,e,i,o,u). More recently, in an effort to help outsiders pronounce Hawai'ian, the glottal stop has been added – marked by ' – (for example, *Hawai'i*, to indicate that each *i* should be pronounced separately).

A horizontal line called a macron is often placed over vowels to be given a longer duration. Technically, each and every letter is pronounced in Hawai'ian, except for a few vowel combinations. However, locals often shorten names a bit, so listen carefully to the way natives pronounce a name. Another addition to the language is a form of pidgin, which served to ease the difficulties of having multiple languages spoken. Laborers were brought in speaking Japanese, Mandarin, Cantonese, Portuguese, English, as well as other languages, and they had to be able to work together. Pidgin evolved as an ad hoc, but surprisingly effective way to communicate, and much of it survives in slang and common usage today.

PRONUNCIATION

Consonants are pronounced the same as in English, except that the W sometimes sounds more like a V when it appears in the middle of a word. Vowels are pronounced as follows:

a = long as in father
e = short as in den, or long as the *ay* in say
i = long as the *ee* in see
o = round as in no
u = round as the *ou* in you

When vowels are joined (as they often are), pronounce each, with slightly more emphasis on the first one. This varies with local usage.

COMMONLY USED VOCABULARY AND PLACE NAMES

a'a = rough lava (of Hawai'ian origin, now used throughout the world)
ahi = yellowfin tuna (albacore)
ahupua'a = land division in pie shape from mountain to sea
ali'i = chief
aloha = hello, goodbye, expressing affection, showing kindness
haole = foreigner (now usually meaning a white person)
heiau = temple, religious platform
hukilau = joining together to pull in the net, then have a party
hula = native Hawai'ian dance
humuhumunukunukuapua'a = trigger fish that is Hawai'ian state fish
imu = pit for steaming food over hot stones
kahuna = powerful priest
kai = sea
kama'aina = long-time resident of the islands
kaola = barbequed
kapu = taboo
kaukau = food
kona = leeward, or away from the direction of the wind
kukui = candlenut (state tree)
lei = garland of flowers, shells, etc. given as a symbol of affection
lu'au = Hawai'ian traditional feast, often including roast pork and poi
mahalo = thanks; admiration, praise, respect
mahi mahi = dolphinfish (not a dolphin)
makai = on the seaside, towards the sea, or in that direction
malihini = recent arrival to the islands, tourist, stranger
mana = power coming from the spirit world
mano = shark
mauka = inland, upland, towards the mountains; in that direction
mauna = mountain, peak
menehune = little people of legend, here before the Polynesians
moana = ocean
nene = Hawai'ian state bird
niu = coconut
'ohana = extended family
'ono = the best, delicious, savory; to relish or crave
pa'hoehoe = lava that has a smooth texture
pakalolo = marijuana
pali = cliff
pu pu = appetizer
taro = starchy rootplant used to make poi
wahine = female
wai = fresh water
wana = black spined sea urchin

TRAVEL TIPS

GETTING THERE

Keahole Airport is located just eleven miles north of Kailua-Kona on Highway 19 (*see map, page 57*). Apart from rush-hour traffic times, it's an easy 15-20 minute drive from town.

Many airlines fly to Honolulu, and then connect to the Big Island. When there are airfare promotions, they tend to be on this route. Most flights used to arrive in Hilo (see map, page 81), but now most of the big jets head for Kona.

Several nonstop flights are available from the mainland, scheduled and charter. From the West Coast, you can fly nonstop in the morning, and be at your condo and in the water by not long after lunch! Plan ahead – the nonstops tend to fill up.

If this is your first trip to the Big Island, don't be fooled by the barren landscape as you land. Your airline hasn't taken you to the moon by accident. The airport is built on lava without the benefit of surrounding golf courses, palms and abundant flowers that flourish elsewhere on the island.

The gate area is nice and small, making it easy to get in and out. This may change in the future, as Kona grows in popularity. Some rental car offices are located directly across the street from the boarding area; the rest have connecting shuttles on this same street.

Parking is still quite casual compared to most airports, so you can usually find a spot directly in front to pick up passengers and luggage. All in all, this is an easy and pleasant arrival and departure – made even more so by the friendly airport employees.

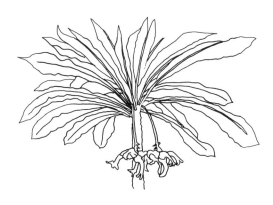

TRAVEL LIGHT

Learn to travel light – most experienced travelers will recommend this. A few good reasons why:

- Traveling carry-on-only eliminates any risk and worry about lost or delayed luggage, which can wreak havoc on your vacation.

- Not having checked luggage makes it easy to catch an earlier connection – just walk on. It also gives you the opportunity to be voluntarily bumped for compensation – which happens often on these heavily booked flights.

- Traveling carry-on only forces you to pack lighter, which is easier on the back when you have to lug everything around.

- Because you don't have to check any bags, there's less standing in line required. It's typical to save an hour or more each way when you have no luggage to check in and out.

LEARNING TO PACK LIGHT

- Develop a wardrobe where everything goes with everything else. Pick a basic color you like – maybe khaki or denim, which won't show dirt so much – then add bright touches with shirts. Polo-type lightweight knit cotton short-sleeve shirts suit the climate well. Casual shirts are popular and easy to find, either in mainland styles, or floral styles developed here. Locals prefer a comfortable look. Even weddings and parties welcome people in either casual or more formal attire. It's a charming Hawai'ian tradition in tune with the tropics.

- Collect travel size (small) shampoo bottles, etc., and refill.

- Use layers. One sweater for the plane will also suffice for all cool circumstances in Hawai'i (unless camping or heading for high altitude). Only in the winter do you need to consider a second sweater or jacket.

- Sandals are acceptable everywhere and easily available at shops and markets. If you don't have any, get them once you arrive, as the variety is probably better than at home. Tennis shoes or hiking boots are necessary for hikes and lava, but this heavy stuff can be worn on the plane if it won't fit in your carry-on.

- Reef shoes and flip-flops are available everywhere, too; however, don't count on easily finding your size in exactly the style you like. Retailers don't necessarily get to choose their sizes, and can end up with lots of odd sizes and colors.

- Make yourself a simple cotton shoe bag, to help keep your bag clean, or use a plastic kitchen trash bag.

- Wear shorts everywhere if you want to blend in (except at very tony resort dining rooms). Even these have been known to make exceptions for paying guests. Long pants are the exception, unless you're working or in a formal situation.

- Use a packable hat, unless you prefer to wear one for the flight. A broad-brim lightweight hat is worth carrying on, if you have one you like.

- It's becoming popular to take a travel vest with lots of pockets for the plane and day trips. This can eliminate the need for a purse or bag, and makes it easy to carry whatever matters to you, such as sunscreen, snacks, folding hairbrush, medicine, reading glasses, a small pen, and so forth.

You really can travel light if you have the discipline – even fins and all snorkeling gear will fit in a standard carry-on. And you're less likely to forget something important (your bathing suit, perhaps?) when you take less.

BUSES AND SHUTTLES

The Hele-On-Bus goes around the island ($5-7) very slowly, but is interesting if you have lots of patience and like seeing local color and watching people. As a local bus, stopping everywhere, it is not quite the same as a commercial round-the-island tour, but it certainly is an excellent value. 961-6722

The Ali'i Shuttle runs between Kona Surf Hotel and Lanihau Shopping Center in Kailua-Kona with stops at major hotels along the way. The red, white and blue-colored bus runs about every 45 minutes ($1 each way). Call or ask first, because times and rates tend to change depending on demand. 322-3500

The Grand Circle Island Tour, $50-60 depending on starting point, per adult, includes garden and volcano park fees, in what they describe as 'a deluxe mini-coach'. 329-1688

VALUABLES

Theft is mostly a problem with rental cars, tourist parking areas and popular beaches or remote beaches – a bit worse on O'ahu because of the larger population, but seen on all the popular islands these days. Of course, urban theft is becoming common in many mainland cities, so why should the Hawai'ian islands be an exception?

Don't get mad – get ready. Follow some easy and sensible precautions. You can greatly decrease your chances of having a problem and also minimize the impact on your vacation if you do get unlucky. Even with our extensive travel in Hawai'i (always with a rental car and often to remote locations) we have never had a single theft. We try not to make ourselves a juicy target, and that seems to help our odds.

It's probably more common to simply lose things – particularly keys. The following ideas are useful for travel to any destination – by no means limited to Hawai'i:

- Leave nothing valuable in your car, not even in the trunk. If what you leave is not valuable, make that obvious. You don't want someone to take your old clothes thinking a wallet might be included. It's best to leave the car unlocked, as a common report in the papers is 'smashed side window', which can cost you more than a theft! On a hot day, it's a luxury to be able to leave the windows down anyway.

- Carry cash (and maybe one credit card) in a concealed pocket. Strap-on travelers zip pouches work well, too. Dive shops often carry various sized water-proof containers in case you'd like to keep your money dry while swimming. These can attach to your wrist or bathing suit.

- Consider the value of prescription glasses and sentimental items. They could be stolen by accident and no one would come out ahead. If you absolutely must have prescription glasses to function, it's always a good idea to have a spare along. After all, they may get lost or broken quite easily.

- Be prepared to replace anything essential. Important medicines, for example, can be carried in duplicate (one container in your room, another in the car).

RACCOON BUTTERFLYFISH

- Hawai'i is not the easiest place for designer fashions, expensive shoes, or other expensive clothing, unless you go to a 'total destination' resort and never venture out. The lava and red dust can take their toll, as do sudden rainstorms and wind. Consider leaving them at home, and you'll lower your theft exposure, too. Take advantage of the casual tradition.

- There's more than one way to have your pocket picked. It doesn't hurt to keep your eye on the rental car agencies themselves. Honest mistakes happen, as do other mishaps. Try to see that they don't list your tank as full, when it's just half-full, or talk you into an unneeded upgrade or insurance that you already have. Inspect the car before leaving, and get them to sign off any dents or scratches.

- Hotels and restaurants have also been known to make mistakes in their favor. At a large resort we were once charged double for what already seemed an overpriced dinner.

- Don't sign anything you haven't read carefully. Most credit card companies are getting increasingly reluctant to retrieve your money once you've signed on the dotted line.

- Pay attention to policies regarding cancellations of bookings for hotels, boats, cars, and tours – these are tending to become stricter.

OFTEN HEARD MYTHS

- **"You'll probably never see a shark."**

  If you snorkel often, you probably will see one eventually, although probably not a Great White or Tiger Shark. Most sharks aren't interested in you for dinner. If you look at actual statistics, your time is better spent worrying about lightning.

- **"Barracudas are harmless to humans."**

  Perhaps some are quite innocuous, but others have bitten off fingers or hands. The Great Barracuda has been involved in the majority of cases we've read. I wouldn't worry about one that has been hanging out in front of a hotel for years, but I wouldn't crowd them either. I'd be even more cautious about eating one for dinner, because they are a definite, major cause of ciguartera "fish poisoning". They are one of the best tasting fish, though, in our experience. Feeling lucky?

- **"Jewelry attracts barracuda bites."**

  I first heard this rumor from a 12-year-old, and it was later reinforced by numerous books. The idea is that the flash will fool a barracuda into attacking. However, we've never heard of a definite case of a woman losing an ear lobe this way, even though I see people swimming and diving with earrings all the time. The same goes for wedding bands. I keep mine on.

- **"There's no snorkeling on the Big Island. Why go there?"**

  You've got to be kidding!!! Whoever started this rumor obviously never made it into the saltwater!

- **"There aren't any sandy beaches on the Big Island."**

  Well, it's a big place and the condos are rarely located on the sandy beaches. Some beaches are small, some are difficult to get to, but there are some large and exceptionally beautiful ones.

- **"The water in Hawai'i is too cold for comfort."**

  It can be pretty cold, especially at certain times of the year, especially if you go in naked; but there is an alternative. Just wear a thin wetsuit and it will feel a lot like the Caribbean. Or you can wait till late summer and give the water a chance to warm up. Don't expect warm water in April.

- **"It rains all the time on the Big Island"**
  "The Big Island is too hot and sunny"
  "It's always windy on the Big Island"

  On the Big Island you can have the climate of your choice. Don't believe everything you read in advertising literature (like hotel brochures) regarding perfect weather. It does vary, there are seasons, and location matters. It just depends on your personal preferences. You may hit a patch of rain, but on the Kona side, it seldom lasts for long. The typical weather report for Kona is: Tonight–fair; Tomorrow, mostly sunny; for the weekend, sunny except for some upslope clouds in the afternoon. The drama of weather is part of the charm of the tropics – enjoy it as it is, rather than expecting it to be exactly as you want.

- **"Octopuses only come out at night."**

  Some types are nocturnal, some not. We've seen lots in Hawai'i quite active during the day. The hard part is spotting them!

SNORKELING THE INTERNET

VISIT OUR WEB PAGE FOR THE LATEST INFO

http://www.wp.com/snorkel_hawaii

The Internet is just beginning to change the way we communicate and gather information. Just because you're laid back while you're snorkeling doesn't mean you have to be out of touch with the latest information. We figure some of you are really wired. You probably brought your laptop along to poolside just for fun. Maybe you don't even have to get out of the water to log on the net, who knows?

We like speed, too, but book writing and publishing is still a slow business. You wouldn't believe how many long, hard hours we spend slaving away, snorkeling and researching, researching and snorkeling some more, in order to produce the little volume you're holding. Maybe a hundred hours of research gets distilled into one little page of maps and text. It makes me sweat just to think about it.

Oh, yeah, some tough job, I hear someone saying. Some folks say that they get no respect; well, we get no sympathy. We can live with that.

To enable you to get the latest corrections and additions between revisions, we have set up a Web page for your benefit. Pull down that left-most menu on your Web browser, and select 'Open Location', or whatever command your browser uses to send you where you want to go. Type in our web site location:

http://www.wp.com/snorkel_hawaii

Hit Enter, and you're there in a flash. We have posted there the many links to Hawai'ian resources on the World Wide Web (WWW), as well as updates to phone numbers, excursions, and many other goodies. There really are a lot of good resources on the Web, and there are more every day. Check it out!

You can also check out our progress on other snorkeling guidebooks. Or you can find out just how to order copies to send to all your friends. A great Christmas or birthday gift, a lot better than another pair of socks for good old Dad! You can even email us using:

indigo@malinowski.com

We'd love to hear what you like or don't like about our books, as well as reports about your experiences snorkeling. If you've found a great snorkeling site anywhere in the world, let us know via e-mail if you can and we'll share some of our favorites, too.

The Web is changing hourly, so the best way to get current links is to go to our Web page, and just click on them! However, here are a few Hawai'ian links that may be helpful, and hopefully haven't moved or changed already:

Hawai'ian Visitors Bureau. Good looking page with lots of links.
http://www.visit.hawaii.org/

Weather stations can be accessed by clicking on a map of the islands.
http://lumahai.soest.hawaii.edu/island.html

University of Hawai'i Meteorology Weather Server. A great site for those with fast connections and curiosity about weather. Satellite pictures, too!
http://lumahai.soest.hawaii.edu/

Other Hawai'i. Lots of links, many of them commercial.
http://www.visit.hawaii.org/lines/main.html

Hawai'ian kids e-zine. Cute and interactive
http://www.cyber-hawaii.com/aloha/

Hawaiian skindiver Magazine. Seems hunter oriented. Very attractive.
http://peacock.com/skindiver/

Native Tongue, Hawai'ian language page. Teaches some basic Hawai'ian. Read, learn and listen, too.
http://www.aloha-hawaii.com/a_speaking.shtml

Big Island Phone directory. Says residential, but it seems to have business listings too.
http://www.travelphone.com/ihawaii/CONTENTS.HTM

Volcano Watch. USGS page with latest volcano forecasts and information. Find out if the lava is still flowing to the sea.
http://www.soest.hawaii.edu/hvo/index.html

INDEX

ABOUT THE AUTHORS

Judy and Mel Malinowski love to snorkel. They have sought out great snorkeling and cultural experiences since the 70's, traveling to 50-some countries from Anguilla to Zanzibar in the process. Hawai'i keeps drawing them back, and eventually they may become kama'aina.

Although they are certified scuba divers, the lightness and freedom of snorkeling keeps it their favorite recreation. They also enjoy surface diving, sailing and windsurfing.

Mel, Judy and their three children have hosted students from New Zealand, Turkey, and Yugoslavia in their home, and helped hundreds of other families enrich their lives through cultural exchange.

Working with exchange students and traveling as much as their businesses allow has encouraged their interest in the study of languages, from Spanish to Thai and Chinese.

Graduates of Stanford University, they live and work in Palo Alto, California.

ORDER FORM

Indigo
Publications
920 Los Robles Avenue
Palo Alto, CA 94306

Please send _____ copies of *Snorkel Hawai'i: The Big Island*

_____ copies of *Snorkel Hawai'i: Maui and Lana'i*
(*Maui and Lana'i* available in early 1997)

I have enclosed a check for the full amount, including shipping

Price: $14.95 per copy (CA residents add local sales tax)
Shipping: $2.35 per order by Priority Mail
                $10.50 Next day air in the USA up to two copies

On-line orders: point your web browser to
                 http://www.wp.com/snorkel_hawaii
                 and follow the ordering instructions posted
                 there. You may order and pay electronically
                 on-line using your First Virtual card.

Mail orders: Send a copy of this form, along with your
                 payment, to:

                 Indigo Publications
                 920 Los Robles Avenue
                 Palo Alto, CA 94306-3127

Fax: (415) 493-3642
E-mail: indigo@malinowski.com

Ship to:

Print or type
clearly, as
this will be
your shipping
label.